scallop season

scallop

On Nantucket, the word is pronounced *scollop*. It has been pronounced this way since 1659. Visitors often smile when first hearing it, knowing how the outside world homogenizes. They are charmed by, perhaps even envious of, Nantucket's tendency to cling quaintly and stubbornly to her independent ways.

No doubt the twenty-six miles of ocean between Nantucket and America have protected and nurtured this tendency. And in truth this stubbornness, this obstinate independence, is hardly considered by islanders a noteworthy trait. It is more like an expected one. Maybe even a necessary one. And is the main reason why, when there remains only one fleet in the entire world harvesting natural bay scallops, it will probably be Nantucket's.

And it will still be pronounced *scollop*.

scallop season

a nantucket chronicle

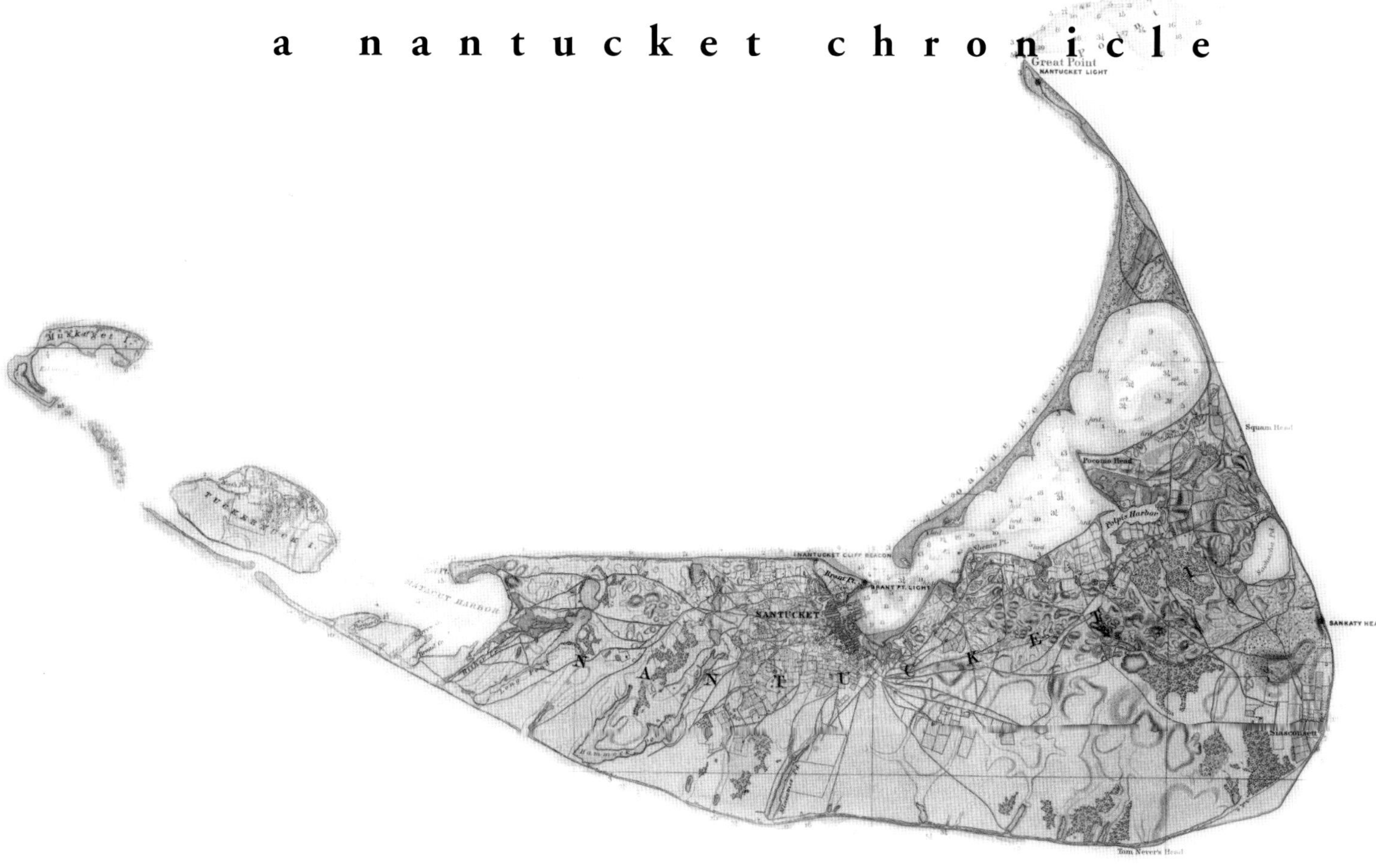

Jim Patrick

Rob Benchley

Autopscot Press

Approach to Nantucket, in 5½ fathoms water, Brant Pt.Light bearing S.by E., distant 1⅞ miles

Scallop Season: A Nantucket Chronicle

To order books, obtain permission to reproduce selections from this book,
see what other books we have in print, or contact us, please visit our website at:
www.autopscot.com, e-mail us at **scallop@autopscot.com**, or write us at: **Autopscot Press, Box 2177, Nantucket, MA 02584**

Book and cover designed by Jim Patrick and Rob Benchley
Text by Jim Patrick, Photography by Rob Benchley
Edited by Martina A. Morrow
Production by Color Graphics, Nantucket

Printed in China
10 9 8 7 6 5 4 3 2 1

Library of Congress Control Number: 2002091379

ISBN: 0-9719615-0-6

The Editor: Martina A. Morrow lives in Portland, Maine and is director at Portland Harbor Museum. She holds a B.A. from Bowdoin College and a Masters in American and New England Studies from the University of Southern Maine. After serving an internship at the Yale University Press and taking a year abroad in the Ukraine and Scotland, Martina worked as docent coordinator for the Nantucket Historical Association and taught at the Nantucket New School.

Acknowledgments

The photographer would like to thank, firstly, all the people who appear in this book; without their trust and assent it would surely be lacking. My apologies to those people I photographed but had to leave out. My wife Carol deserves an enormous amount of credit for carrying me through this project. Not only on an emotional basis, but for two years' worth of photo and computer clutter piled throughout the house. And to Rose, who lost out on many walks during the process, but who made it onto Page 249. Finally, I'd like to thank Da, who gave me my first mini-darkroom kit almost forty years ago. I only wish you could see this.

The map of Nantucket and harbor approach on the previous page are details from a Coast and Geodetic Survey chart, first published in 1860. The Department of Commerce reprinted these beautifully etched charts from the original copper plates in 1957 to celebrate the Survey's 150th anniversary. The 1957 chart reproduced here is from the collection of Mr. and Mrs. William P. Herbert.

--RB

I was fortunate enough to have known Martina Morrow for most of the three years that she lived on Nantucket, teaching, working for the Nantucket Historical Association, and participating in local theatre. She has added much insight, wisdom, knowledge, and guidance to every phase of this book. I, of course, wish to thank her for the many grammatical, factual, and technical errors she caught, but without the very large overall contributions she made, this book would be vastly different and vastly inferior. But mostly, I want to thank her for her continuing friendship and encouragement. I miss her, and the theatre misses her, too.

Many others directly contributed, most importantly the people I interviewed or spoke to, and whose conversations were incorporated into the book. Others, too numerous to name, happily regaled me with their own scalloping experiences, and though their names may not have appeared, pieces of their stories might have worked their way into the text, and certainly added important flavor. All the people who gave me information, knowledge, and history deserve credit for their contributions to the final product, but I hold nobody but myself responsible for any mistakes, errors, or misinterpretations that might have occurred.

I would also like to thank Marianne Stanton, Publisher and Editor of *The Inquirer and Mirror*, for permission to use the editorial 'A Healthy Harbor First' under the chapter of that name, as well as information and quotes derived from that newspaper. Epilogue relies heavily on their news accounts, particularly any individuals who were quoted. Fish House and On The Water were originally published in the *Nantuvket Magazine*.

Part of what keeps a community alive are its gathering places and its sense of hospitality. Over the two year period of writing this book my favorite place was the soda fountain at Nantucket Pharmacy, where Pete and his gang (Claire, Tricia, Mairéad, Kelly, Fallyne, Debbie, Matt, and of course Patti) not only kept me in coffee and muffins, but also unfailingly provided a cheerful and welcoming atmosphere, whether talking about scalloping or talking about life. And whenever I needed to see someone, I knew they would eventually wander in.

Some people whose encouragement, advice, or specific knowledge and assistance have proven immensely valuable are Mimi Beman, whose Mitchell's Book Corner is a true treasure to island readers and authors; Bruce Marshall-Jones at Color Graphics Studio; Max Berry who understands the importance of history to a community; Tom, Gail, and Bill Potter for the earliest critiques - likewise Terri Senft and Nicki Friedmann; Pierre and Gail Ratté, and Mark and Christina Weber for later valuable insights; Jacqueline Egan, whose longtime moral support was backed up with a computer at a crucial juncture; Jenny Lyn Bader who was very generous in helping me understand the publishing process; Emily Levy for both artistic advice and friendship, and Carol Benchley for cheerfully putting up with the many invasions of her living room and endless telephone calls, and who has become a friend in this process. And Rosalie, Patrick, Teresa, and Tina Arthur were in my thoughts always, from the first page to the last.

-- JP

preface

117,000 bushels of scallops were landed in Nantucket waters during the 1980-81 season. 6,800 bushels were landed in 1998-99. While this was not why I wrote this book, it was, however, why I wrote it this year.

The truth is, I had been intrigued by scalloping since my first full winter on Nantucket, 1982-83. It seemed a mysterious, cold, dangerous occupation - physically harsh, yet lush with primeval smells and sounds, tastes and textures. These sensations represented more than just a new experience to me, they represented vicarious thrill. Who were these mysterious people, who live on this tiny, isolated island, twenty-six miles beyond the farthest reach of the continent, whose small boats dot the harbor on bone chilling winter days, who return wet and cold and triumphant, boxes overflowing with scallops, who then vanish into a landscape of shingles and pine trees and meandering dirt roads as though dream figures? I wanted to follow and chronicle them, portray their lives for one full season, beginning to end. But I had soon found my own place in Nantucket's working rhythm, hurrying to finish my various projects before winter set in. Opening day would come and go, and I would put it off until next year, and then the next year, and then the next.

The realization that there might be no next year ended all procrastination. The following day I went to the library to research, and kept running across beautiful, expressive newspaper photographs of scallopers and scallop boats, invariably with Rob Benchley's credit below them. I called Rob that night and met him in town a day later to discuss the idea of the book. We were in agreement before lunch arrived, and at work the next day - he with camera and dog, me with tape recorder and legal pad.

Our approach was to chronicle the season as it unfolded, with fresh eyes and ears, no preconceived notions, and very little foreknowledge. It was quickly evident that seemingly everyone on the island had some connection to scalloping in their life, and the thread of community soon began to weave its own tale. While this book began as the story of scallopers, the scallop industry, and the Nantucket waters that support it, it grew into a story of the community that is nourished both economically and spiritually by it, and is currently engaged in trying to save it.

Jim Patrick
Nantucket, Massachusetts
April 1, 2000

The Chronicle

Scallop Season

The Chronicle

The Chronicle

Scallop Season

Introduction

a thread to the past

Welcome to the Website devoted to preserving the last surviving bay scallop fishery in America.

This is the greeting on the home page of the Nantucket Harbor and Shellfish Advisory Board. It is a reminder that precious few 'lasts' remain, and that many things that once were 'last' are now simply gone. Time passes, new firsts replace old lasts, a new generation inherits the earth, and old lasts are missed no longer.

Roughly four hundred scallop species exist, and perhaps fifty are fished commercially. Nantucket fishermen catch *argopecten irradians irradians*, the northern bay scallop. Like wines, scallops acquire their taste by type and location, and some local varieties are rare and highly prized. Famous for their sweetness and size, Nantucket Bay Scallops are considered by many the finest in the world. They appear on the menus of some of the most exclusive restaurants in the country, often flown in through special arrangement with a seafood distributor, or even the fisherman himself. But, while Nantucket's scallops have always ranked among the best, bay scallops in general were not always such a rarity. They were once so plentiful that they were used as fertilizer. But as of this writing, they support commercial fleets only on the Massachusetts islands of Nantucket and Martha's Vineyard, off the south coast of Cape Cod, and occasionally provide a year's catch to fishermen in a handful of other Northeast fishing towns.

A 1931 eel grass blight destroyed most northern East Coast fisheries. Those that rebuilt, chief among them Long Island's famous Peconic Bay, have twice since been decimated by recurring brown tides. Some have tried to rebuild through propagation efforts, most have given up. The vast majority of commercial bay scallops are now farmed in China or Chile, with some natural Nova Scotia Digbys, Icelandic or Florida Calicos, Mexican Bajas, or Carolina southern bays contributing to the catch. Sea scallops have long since replaced bay scallops as the major commercial catch; they are typically frozen and shipped worldwide.

Bay scallops were first caught commercially in the latter part of the 1800's in the many protected, shallow-water bays and estuaries of the Atlantic seaboard. Initially, scallop fishermen waded out with pushrakes or nets, or collected them by hand. By the turn of the century, most scalloping was being done from small, shallow-draft day boats, towing dredges. It eventually became an important source of winter income to many fishermen, who tend to rotate catch types seasonally. Naturally harvested bay scallops dominated the market late into the 20th century.

The story of the Nantucket scallop fishery paralleled that of its East Coast cousins until the 1931 eel grass blight, which Nantucket managed to survive better than most. Although many fisheries never caught another scallop, a few did manage to rebuild their stocks. In fact, as late as the 1981-82 season, the great Peconic Bay was harvesting scallops in numbers that rivaled its best years ever. And on Nantucket, over 200 boats had been catching their limits all winter long for at least three decades. But in the mid-1980's another mysterious eel grass blight devastated the Peconic, and seemed to affect the few Connecticut, Rhode Island, and Cape Cod harbors that still supported modest fleets. In the mid-1990's, just as they were rebounding, another blight struck. This was a virtual death blow. By 1995, only Nantucket and Martha's Vineyard supported fleets, and the Vineyard was aided by a massive, state-of-the-art seeding program they had begun twenty years earlier. And by the 1998-99 season, after a precipitous three-year decline, the Nantucket catch had reached its all time low. Even so, 134 permits were sold and roughly 70 boats planned to fish in 1999-2000.

How has the Nantucket fishery survived this far? Partly, Nantucket's physical isolation from the mainland helped preserve its pristine beauty and natural resources. And partly, for nearly a century, Nantucket remained an exclusive and remote summer destination for the discreetly wealthy. These factors created a hierarchy of power, based on the intertwining and often intangible relationships of history and culture, family and wealth, skill and temperament, and status as native or off-islander. This complex dynamic has long accorded tradesmen and tycoons roughly equal social standing, dating back to its early social system of proprietors and half-sharemen. It was part of the same dynamic of an emerging 'American' consciousness that was sowing seeds of revolution throughout the young colonies at that time, and that would flower nearly a century later.

This physical distance from mainland soil, and spiritual distance from mainland thinking, has helped bond people of disparate backgrounds into a single term, 'islander', and has contributed heavily to Nantucket's ability to maintain its scallop fishery as a viable industry and an important cultural heritage. Nantucket's fishery is far more than a business. It is a direct thread to its past.

an unfolding season

Scallop season on Nantucket means two entirely different things: family scalloping and commercial scalloping.

Family scalloping, or recreational scalloping as it is properly called, begins on October 1st and runs to March 31st. Family scallopers can gather one bushel per week, using manual techniques only, such as pushrakes or dipnets, while wading, diving, or snorkeling. It is called family season because these scalloping adventures are ideal times for families to gather and reunite over the common task of collecting, opening, and cooking the prized shellfish, and symbolically putting the summer tourist season to rest. Most people go on opening day or within the first two or three weeks, when the weather is still sunny and the water is still warm. This year, 1,800 family permits were sold at $15.00 apiece.

Commercial scalloping begins on November 1st and lasts through March 31st. Permits must be applied for by March 31st of the previous year, and cost $250.00. Commercial scallopers typically fish in shallow water from small day boats in Nantucket Harbor, Madaket Harbor, or off the neighboring island of Tuckernuck, towing several dredges behind their boats; the standard is eight. They can use power equipment and collect five bushels per day per permit, with two permits allowed per boat. Commercial scalloping is done largely during the late fall and winter months, and is cold, difficult work.

Rob Benchley and I followed Nantucket's commercial scalloping community through what was in many ways a typical season. We went with scallopers on their boats, met them at the docks as they arrived, talked with them at their homes or workplaces, or at their favorite breakfast or lunch spots. We also visited 'shuckers' who open the scallops, buyers at the fish markets, wardens who try to catch scallopers doing something wrong, and town officials responsible for regulating and maintaining the fishery and the harbors.

While this book focuses primarily on the commercial fishery, the first few pages offer a brief overview of family scalloping as well. After that, the 1999-2000 season unfolds for the reader much the way as it unfolded for Rob and myself.

October, 1999

nine scallops

October is reserved for recreational scallopers, generally known as family scallopers. The first week of October is absolutely sacrosanct in many Nantucket families - both year round and summer resident alike. Reunions and house closings and winter preparations happen this week, many centered around that first opportunity to catch one of the world's great delicacies.

Family scallopers are often doubly blessed with glorious summery weather and an island resplendent in the colors of early fall, as they wade and splash and try various methods of pushraking, dipnetting, diving, snorkeling, or simply reaching down and grabbing, hoping to catch their weekly limit of one bushel.

People who live on the island wait all summer for opening day, and people who live off-island plan their return year after year. The ritual often dates back three or more generations, and missing opening day is an offense surpassed in magnitude only by revealing the family's favorite fishing spot.

Henry Mitchell and Mary Allen Havemeyer came to appreciate this ritual, and to understand the true nature of the Nantucketer in regards to scalloping, shortly after opening their store, Mitchell's Book Corner, in June of 1968. Mitch, a retired ad man, had run weekly book descriptions throughout the summer. But when October came, he broke out of the mold in more ways than one. He and Mary Allen went scalloping, this famous and beloved enterprise they had heard so much about, and had recently read about in Euell Gibbons' soon-to-be legendary *Stalking the Blue-Eyed Scallop*, and another book whose author was as quickly forgotten called *Catch 'em and Cook 'em*. The Havemeyers, in one of the best scalloping years ever, landed not quite as many scallops as Mary Allen could count on two hands, and Mitch's next ad, cleverly plugging the books while detailing their misadventure, was titled 'We Caught Nine Scallops'. The day the ad ran, they returned home to find scallops in their freezer, and equipment in their yard. It was the first of many delightful scalloping stories Mitch would write. My favorite was written one year later, entitled *The Noble Scalloper*.

The Noble Scalloper

Even nobler than the Great Nantucket Bay Scallop is the man who hunts him.

Witness his fortitude. Having squeezed himself into his scallopsuit, he ventures seaward. Nonchalantly he secures to his midsection a short painter. Attached and floating a yard away is a plump, black inner tube carrying a basket ready to receive his catch. With great dignity he starts pushing his net (or pulling his rake).

The harvest of "grass" is excellent, and stones are abundant. This year, however, the Noble Scallop, for reasons best known to himself, has mostly jetted off to parts unknown. The few that have remained as a rear guard must have been selected for their strength and defiance. At every opportunity, while hiding in the dripping grass, they ambush one's fingers. Although a scallop bite is rarely, if ever, fatal, it is decidedly harmful to one's aplomb.

While all this is going on, the scalloper, now waist-deep in the Harbor, discovers that his waders' right leg is not entirely waterproof. Slowly, inexorably, the level of water rises. Two bushels of grass (and four scallops) later the ice water reaches the top of the midships bulkhead and starts trickling down the left leg. Suddenly reminded of the "Titanic", our scallop hunter exclaims to himself, "Abandon ship!" Which, in turn, reminds him to rescue his wallet from his left hip pocket. And just in time.

The rising tide and impending frostbite of the lower limbs forces our Noble Scalloper to head for shore - with his ridiculous contraption in tow. By now the basket has imprisoned a couple of hundred vicious, snarling scallops - almost as dangerous as they are delicious. In fact many people consider them the most delectable of all foods.

The advertisement, by the way, was for a best seller about the *Titanic* called *Maiden Voyage*.

Gathering food from nature, from the sea, is one of the most enduring of human instincts. It may be a reason why we return to it year after year.

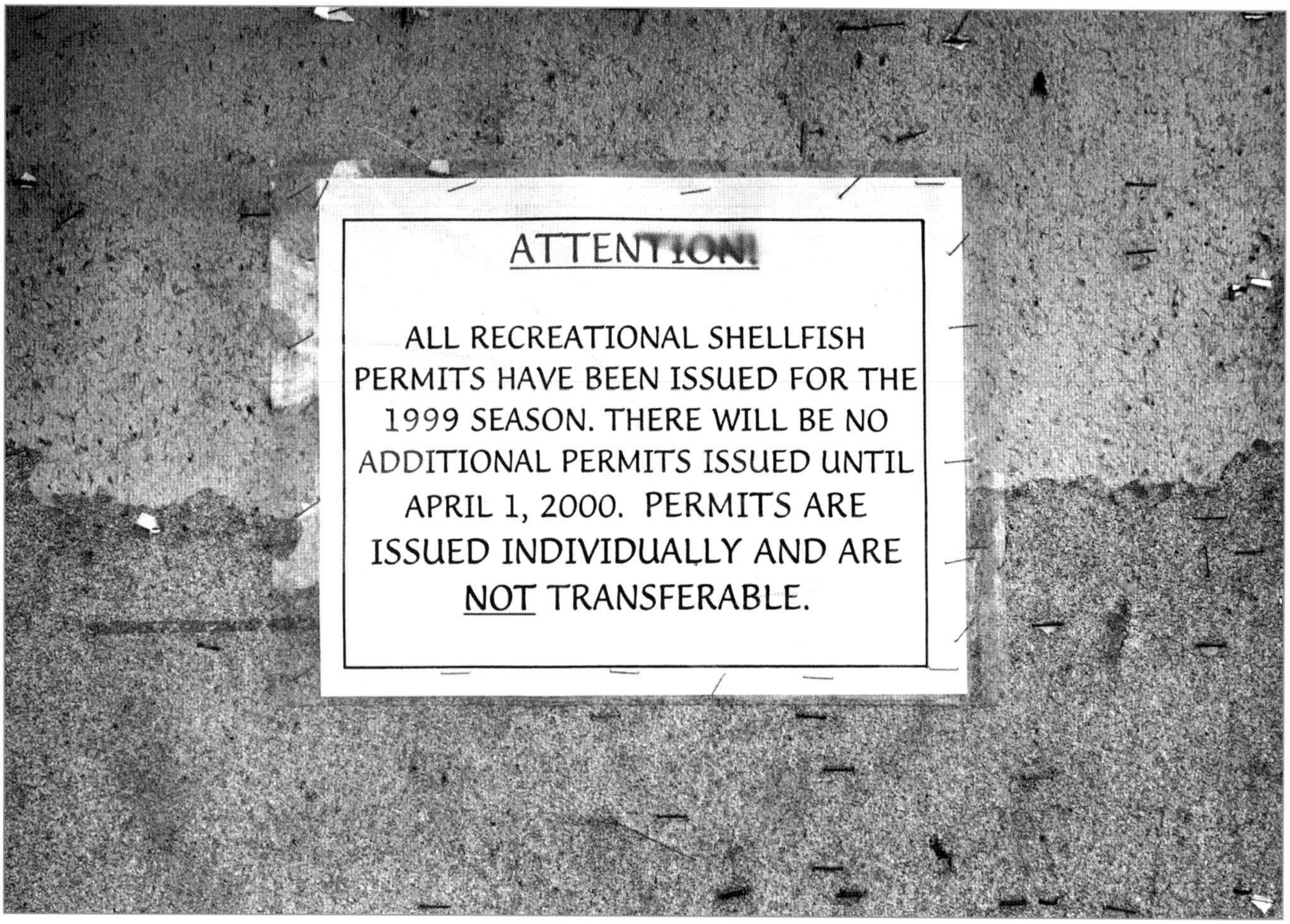

For the first time ever, the number of so-called 'family permits' was capped at 1,800. Demand was high because it has been such an unexpectedly good scallop year in Madaket. A recreational permit costs fifteen dollars. Commercial permits cost two hundred and fifty dollars, and must be bought by March 31st for the following season.

Teresa and Cristina Arthur caught their first scallops in Madaket this year. They got their shirts at Brant Point Marine.

A recreational license allows one bushel per week for personal use, beginning October 1. A dog license allows an early autumn swim.

Equipment includes pushrakes to gather them in and a basket suspended in an innertube to throw your keepers in. Waders and wetsuits keep scallopers dry and warm, but on a sunny October day, one can work up a good sweat.

Katherine and Rob Baker, nearing their limit in Madaket

benchley's button

Rob told me about a tradition he would like to begin. In his own words:

Gerry Eldridge has a lot to be proud of. For one, he has been Superintendent of the Siasconset Water Company for 23 years; secondly, he is proud of his father, Kenneth, who ran the same water company 25 years before him. For the better part of fifty years, Jerry and Kenneth have supplied 'Sconset with pure water for all its various uses: fire protection, dishwashers, toilets, kitchen sinks, lawn sprinklers, drinking fountains and ice cubes.

And Gerry has another particular pride. Year after year he has somehow managed to obtain the coveted lowest number on the annual family shellfish permit. And year after year, Gerry loves to quietly show off his badge of honor: Number '1'. It remains a mystery how he lands Number '1' every year. If anyone asks, Gerry will happily flash his badge; but he's coy about saying how he got it.

This family permit, issued by the Town by way of the Marine Department, is one of the last best deals on the island. For $15.00, you get a colored pin-on button (the color changes each year to keep everyone honest) and a printed guide of the shellfishing regulations. This unassuming little badge opens a door to the delectable galaxy of gathering your own steamers, quahogs, mussels, oysters and (if you like) eels. And of course, the bay scallop. Gerry puts Number '1' to good use throughout the fall and winter, collecting steamers for his famous 'Portuguese Boil', a heavenly combination of clams, onions, potatoes, baby linguica sausage and a few other secret ingredients.

I go to the Marine Department good and early every April and rifle through the basketful of badges, but the lowest number I could ever get was #12. I have saved every button since I have started, and they make a colorful collage in that special drawer reserved only for old family scalloping buttons. But my collection will not be complete until I have the one that says it all. Number '1'.

Into early adulthood, I summered on Nantucket with my parents and all manner of aunts, uncles, cousins and grandparents. In those '50s and '60s it was a kid's world of harbor exploration and all the unfolding wonders that were to be discovered just beneath the surface. When we weren't at the South Shore fishing or getting boiled in the surf, we were shuffling around Polpis Harbor and the lagoon at Great Point for quahogs and mussels, and we netted blue crabs from the bridge over Hither Creek. Many good meals came out of our efforts. But since we were here only during the summer, little did we know that we were missing the magical 'other season', September through May, the season when the family permit allows one to gather oysters, soft-shell clams, and the bay scallop.

I was first infected by the scallop bug almost 25 years ago during an October visit with brothers Mark and Peter Fredland. We found Polpis Harbor loaded with them, and I had my very first bushel of scallops in half an hour, which was followed by my very first shucking lesson. It was a delightful experience and the ensuing dinners were memorable.

Back in Boston, I realized I was weary from my work and the growing crime

The Collection

in my neighborhood, so I held my breath, landed a job at *The Inquirer and Mirror* and made the move back to Nantucket; you could almost say that it was the scallops that made me do it.

Gearing up for this family scalloping affliction begins with either buying or borrowing chest waders, basket and inner tube, a push rake or a floating viewing box, and a dip net.

The push rake is the most common device for gathering scallops. (The name itself evokes one of those colorful Island idioms. How do you push a rake?) It is a simple tool made of a ten-to-fourteen foot wooden pole attached to a rectangular iron frame on which is hung a pouch of heavy fish net. The family scalloper pushes one end of the pole, the frame bumps along the bottom or through the eel grass, and the scallops flop into the net. This technique takes a lot of faith, since you can't actually see the scallops, and it is not easy work. The rake hangs up in the grass, snags on rocks, and digs into the sand. But a longer pole allows a lower angle of attack, increasing the prospect of netting one scallop or twenty, if there are any down there in the first place.

A second method involves the viewing box, an uncomplicated appliance made of a square wooden frame with a glass bottom. The scalloper hunches over his floating box, spies the quarry on the bottom, and scoops it up with a dip net. It's a pleasant way of going because you can see what you're after, and there is so much other life to see down there. Both the push rake and the dip net methods embrace the same peril, however. Just one step deeper than the top of your waders awaits the rushing-in of the harbor. It's a lot easier to do than you'd think, a lot colder than you'd think as well.

Several seasons of this shivering experience led me to a third method used by Mark and Peter, which is to get wet in the first instance. I now slip into a full neoprene wetsuit with gloves, hood, and booties, and with snorkel and mask I simply jump in. The first immersion is a bit shocking, but in less than a minute the cold water between the skin and the suit warms up nicely.

Opening Day, October First, is usually one glorious day. It is regularly warm and still, soaked in the soft light of early fall and full of happy anticipation. I have not missed Opening Day since 1983. And as the category of the license declares, it's a family thing. My aunt and uncle, Bill and Mary Herbert, have joined in, and we've pretty much settled on the wetsuit method. Actually, Uncle Bill does the scalloping; Aunt Mary always comes along for the ride and the companionship, and to catch up on her knitting. We mark our calendars sometime in December to make sure that nothing else is planned for Opening Day. We then occupy ourselves for the next ten months, messing about with less important things until October First.

Uncle Bill and Aunt Mary Herbert at Madaket, Opening Day.

Brother George culls at Monomoy

At first, the term 'family scalloping' seemed just a way to distinguish the activity from 'commercial scalloping.' I didn't find much 'family' to it. In fact, it seemed a good way to get away from the family. Snorkeling, in particular, is rather solitary work. It's just you and the underwater world. In fact, I once renamed my family permit the 'hermit permit.' But as family started fading away, and perhaps I started growing up some, family scalloping evolved into just that: a no-stress-allowed zone where a family, however one defines it, can get re-connected and work toward a common cause. Even the family dog can get into the act. A scallop dinner cannot be such a bad goal.

In the early days we snorkeled in Polpis Harbor where the scallops were abundant, as were other surprises in those sheltered waters. With the snorkel/mask/dip net method, we could chase down the fierce (and tasty) blue crab. I once netted a flounder the size of a manhole cover just ten feet from shore. I put the dip net in front of him, tapped on his tail with my foot, and into the net he swam. I was so surprised at how ridiculously easy this was, I let him go.

Another time while prowling around Polpis I came upon something the color of the bottom, hiding in the eel grass. This ugly beast was all head and mouth with the face of contempt - as my father would have said - "A face you'd like to slap". When I poked it with my dip net, it lunged at me, hideous, tooth-filled mouth all agape. Screaming through my snorkel, bubbles flying, arms flailing, I backpedaled through eel grass until I was safely out of the water. I never poked at an ugly fish again.

Later that year, on a windless January day I felt a wave of the scallop bug. I went to my usual Polpis haunt and donned two wetsuits; a shorty underneath and a full arms-and-legs suit over the top, plus gloves, booties and a hood. Hammers hit the skin as I eased myself into the 35-degree water. Any illusions about the water not being cold were laid to rest by the fact that most of the harbor was covered with ice. No problem, because back then we always administered two shots of Jim Beam while suiting up, for insulation purposes only.

Underwater I saw a few scallops here and there, but then spotted a nice cluster out of reach in deep water. Letting go of the tether to my inner tube, I took a deep breath, dove down, got five or six in my net, and popped back up to the surface. But I didn't find the surface, I found the underside of the ice. I clawed and pounded at it, and the word 'stupid' came to mind. In full panic now, my hand happened to snag the line to the inner tube, and I was able to yank myself to safety. All this for a few scallops? Maybe.

I learned three lessons that day: First: family scalloping means 'don't go by yourself,' especially in the dead of winter when nobody is around. Second: stay away from ice. And third: stay away from bourbon - it makes a person stupid.

In those early days, when we loaded up our baskets and our freezers with scallops, we were all heady with the activity and the plenty. But in the mid-1980's we noticed a change beginning, especially in Polpis Harbor. The eel grass seemed to be disappearing, and an invasive, non-native, thick-bodied seaweed called codium was choking the lee shores. In two or three seasons the scallops just disappeared. The quahogs had dropped off too, and there were fewer and fewer oysters as well. Many of the blue crabs were dead ones. One thing there seemed to be more and more of were conch, oyster drills and green crabs. Within a few years, the bottom was practically bare, and to the eye, lifeless.

We moved to other places. Uncle Billy and Aunt Mary and I still always check Pocomo, on both sides of the point. Some years the scallop beds were really thick, other years they were practically barren. Three years ago we turned up a goose-egg, last year about 25 people joined us working the ground pretty hard on Opening Day, but we were mostly finding big seed. Monomoy was great one year, lousy the next. And just when we'd think it was about to collapse, a hot spot would arise: in 1999, Jackson's Point was so loaded with scallops they crunched underfoot, and we got our limit in 15 minutes.

These days we're just as excited about scalloping, but less frenetic. When we find them, we take just what we need for a couple of meals. Besides, if you catch a bushel of scallops, you have to open a bushel of scallops.

Meanwhile, Gerry Eldridge has hinted recently that he plans to retire from the 'Sconset Water Company, but he worries about leaving the water plant in good hands (as ice cubes are very important in the village). With his retirement, he'll be eligible for the White Button; a permanent, as-long-as-you-live, free-of-charge, family shellfish license.

Which also means maybe now he'll tell me how to get that badge: Number '1'.

Polpis harbor, one of the family haunts.

gearing up

During the last week of October, the commercial fleet is in full swing, gearing up for their opening day. November first will fall on a Monday this year, so the final weekend is bustling with activity. Back yards, side yards, front yards, and driveways are littered with hulls being painted, dredges being repaired, and engines being tuned. Lines are being laid out and checked for fraying and hard weather gear is being examined for wear and tear. Shanties are being hosed down, bleached, painted white, and fiberglassed. A steady stream of trailers clogs the landing areas, as those scallopers whose boats are ready launch them.

One hundred and thirty four permits were issued this year, and there will be around seventy boats on the water. Nantucket has the largest bay scallop fleet in the country. Scallopers cross paths at Marine Home Center or Island Lumber or Brant Point Marine. They wander to their favorite landing places and gaze out over the water to their moorings as though to reassure themselves that neither had somehow vanished over the summer. They glance toward their favorite fishing grounds and hope to keep them secret again this year. Pick-up trucks stop in the street and drivers call out phrases like - "Frankie! Fishin' out of that old tub again?", or "Better keep on painting - only thing keepin' 'er afloat!", or the time-honored "Missed a spot!" Other trucks honk and their drivers shout epithets about 'Damn tourists holding up traffic', or 'Not everybody's on vacation you know!' More scallopers on their way home from picking up supplies and looking at the water. The same four questions are repeated dozens of times a day: "How's the boat coming?" "Gonna be a good season?" "You fishin' Madaket or town?" "Heard anything about the price?" And, with minor variations, the same replies are given: "Just about make it, any luck." "Better than last year, I hope." "I'm fishin' wherever they are, where you fishing?" "Find out Monday, I guess."

And though these good-natured scallopers with their arms hanging out of pick-up doors, clustered like teenagers on streetcorners, loitering around anyone working on a boat, might not seem too outwardly impatient or excited, they are. And they are all waiting for the same moment: 6:30 a.m., Monday morning, when they can drop their dredges in the water and start scalloping again.

king of the fleet

In June of 1956, a young man made his first trip in his own scallop boat: a 1953 Chris-Craft which had been built by Albert Ottison and Lester 'Buddy' Ayers from a Sears kit. He had paid Lester four hundred dollars down for the two year old boat, plus twelve more payments of fifty dollars, and considered himself honored. Lester was, in his eyes, the best scalloper on the water.

Bringing the boat from Madaket into Nantucket Harbor, he went around Eel Point and across the Dionis shore. As he entered the channel between the Jetties, the tide was running out and a strong north wind was blowing in - ideal conditions to build up waves. The young man was no stranger to a scallop boat or to the water. He had started scalloping as a teenager five years earlier with his stepfather and stepbrother, in a homemade skiff that had cost his stepfather all of seventy-five dollars. The young man saved until he could afford to buy his own boat, and was now bringing it into the harbor. But, with the stubborn current rushing out and the stiff northerly breeze continuing to pile up waves against his stern, he felt the first pangs of trouble. Each swell pushed the boat along, sometimes toward the port side and back into the channel, sometimes off to starboard and towards the boulders that form the West Jetty. It is not unheard of to founder on the Jetties in even calmer seas. But, attempting to turn around and go back in such conditions leaves one momentarily vulnerable to being swamped broadside, or capsized.

The young man persevered, maneuvered his 1953 Chris-Craft into the harbor, and fished with it the following November. He will be fishing in the same boat this year, and if there are still scallops to be caught, he will be fishing next year as well.

Eddie Stojak is the young man who guided his new boat on that harrowing trip through the Jetties forty-three years ago. And he still has the same 1953 Chris-Craft. It is now the oldest boat in the fleet. There is a famous hedge at the top of Cobblestone Hill, overlooking Jetties beach, which was owned for years by the Reighley family, and is now home to a world famous clothing designer. It rises perhaps twelve feet above a mottled and mossy concrete retaining wall which angles up a steep hill. The hedge emerges so straight and square from the top of the wall that, at night, it can be mistaken as an extension of it. Eddie has trimmed that hedge for forty-six years with the precision of a carpenter with a level. He brings the same energy and fastidiousness to every aspect of his life, scalloping included. His shucking shanty was used as a model when regulations on the building and cleanliness of shanties were first introduced. A television tucked in the

Eddie (Herky) Stojak, King of the Fleet.

upper-right corner even adds a homey touch. Eddie's wife Shirley often helps him with the monotonous work of opening the day's catch.

Today, on a Sunday afternoon, about a week before the commercial season begins, Eddie Stojak stands proudly next to his shanty and beside his boat, which gleams under a bright sun and a crisp sky. The smell of Benjamin Moore's Platinum Gray, applied yesterday to the trim and rails, hangs delicately in today's light morning air. The blue marine paint below the waterline is still fresh, while the main hull had been painted white some days before.

Eddie came out of the hospital for kidney stone surgery Friday night, and was out working on his boat Saturday morning. He would be thinking about retirement if it weren't for the fact that he likes to work. Most scallopers do. He has fished hard for fifty years now and run a landscaping business nearly as long. Any Nantucket scalloper will tell you that Eddie Stojak is the King of the Fleet. Any scalloper, that is, besides Eddie himself, who tends toward Yankee matter-of-fact modesty when it comes to himself. He'll tell you about Buddy Ayers - "He was top, he was the best around." Or a couple of others of that era, like Billy Ray Reis or Franky Murray. "They were really hard to keep up with," says Eddie. "They could really move."

Eddie started scalloping around 1950, while he was still in high school. He would go on Saturdays with his stepfather and stepbrother, Joseph Viera and Joe Jr., who was a little older than Eddie. Meticulous upkeep has kept the old Chris-Craft in good shape all these years, and only a few modifications have been made to her since she was assembled. Eddie tried a fiberglass culling board one year, but never liked it - now a simple piece of plywood suffices. "I had to replace a little up near the bow about thirty years ago," he says. "Pushed through quite a bit of ice that year." He tells me that winters used to be much colder. He remembers fishing near Oscar Bunting one year when the ice was so high, piled up by tides and currents, that Oscar's boat disappeared completely behind one of the floes.

This year's project was replacing the stern, ripping out both of the old pieces, and screwing the new piece of plywood into place, then fiberglassing around the back and the bottom, and finally sanding down and painting. Eddie has removed the old transom and replaced it with three-quarter inch plywood, notched to receive his outboard engine, and shaped precisely to fit the pattern of the hull, which is all three-eighths inch mahogany plywood. He picks up two pieces of worn wood and explains that one was the original stern, similarly notched and shaped, and reinforced with half-inch plywood from when he first converted the boat from an inboard to an outboard. Two old-timers told him he'd ruined a perfectly good boat, but he knew he needed an outboard to fish the shallows, and it has served him well.

Eddie's craftsmanship is superb. Except for the nearby scrap pile, one would never know that any work had been done. He makes a point of mentioning that he has given his boat a fresh coat of paint every year for forty-three years.

"Not many boats last that long," he says. "Not work boats." And then he adds, with just a hint of pride sneaking out from under a self-deprecating grin, " 'Course I'm the only guy who'd be doing this. Most guys would have hauled her off to the dump by now."

overlooking the harbor

Fourteen hats hang around Dave Fronzuto's window and he wears them all. He is Nantucket's Harbormaster. From his office at the foot of town pier, he oversees virtually everything relating to salt water around Nantucket. If you want to catch an eel, you see him for a permit. If you want to moor your eighty-five foot yacht, you see him for a slip. He hires the lifeguards, and he hires the wardens. He patrols the harbor for boaters in distress and boaters illegally discharging wastewater. His input is requested in the beach access controversy, the piping plover protection debate, the planning of new mooring fields, a new floating pier, and the planning of a new storm drain system. His office issues permits and regulates every kind of fishing, kayak rentals, windsurfer rentals, boat rentals, commercial fishing and charter boating, and the much-valued commercial scalloping licenses. But his purview extends far beyond the issuing of permits and enforcement of minor infringements.

On the Wednesday before opening day of the commercial scalloping season,

I walk unannounced into his second story office, which commands a view of town pier and the long stretch of Nantucket Harbor. I have picked a bad time. Dave is meeting with Tracy Curley, the town biologist who doubles as a shellfish warden during scallop season, and there are a number of pressing issues to be addressed before opening day. Fronzuto instantly refers me to several recent articles on scalloping in *The Inquirer and Mirror,* the *Nantucket Map and Legend,* the *Cape Cod Times,* and *Nantucket Magazine.* He summarizes their relative merits, hands me a videotape the local TV station had made of him discussing the harbor, and indicates again that, although he'd love to accommodate me, now is not a good time. During this three or four minute stretch, two people have yelled up the stairs with one question or another, and the voice of a third in the office below prompts him to yell down in his raspy, growlish voice to a deputy shellfish warden: "Don't forget, you have to get sworn in by Monday!" Then - "Yeah, give him a sticker," he responds to a question from office manager Sue Mueller. Finally - "I really can't talk to you today," he again protests.

Nonetheless, he continues talking, and soon Tracy, who looks a little like Judy Collins in red plaid workman's shirt and jeans, joins in. Five minutes pass, then ten, then fifteen, and eventually I am offered a seat. Charts and graphs and studies are produced, statistics cited, information poured out. Harbormaster and biologist take turns elaborating, covering ground ranging from scallop anatomy to genetic diversity to enforcement procedures.

It is clear that these two individuals are passionate about Nantucket's harbor and its scallop fishery, even if they are not entirely in agreement on what the harbor and the fishery needs to be saved from, or how that will be accomplished.

"Do you know," Dave says, thrusting an unlit Frisco cigar intimidatingly at my chest, "In the last eight years we spent three million dollars on scallop propagation? We've been studied by Woods Hole, we have our own town biologist, we have sixteen grow-out zones working, we have a water quality study coming out in winter and a computer modeling program due in spring. We're the most studied harbor in the world and what have they concluded? Everything we've done is wrong."

I learn more in these few minutes, which eventually stretch into a full half-hour, than I ever imagined there is to know about a simple shellfish. Bay scallops are the most highly prized scallop in the world. Once plentiful in the gentle, shallow bays from northern New Jersey to Cape Cod, Martha's Vineyard and Nantucket now account for the vast majority of bay scallops. Bay scallops have a two year life cycle, generally breeding only once. Their delicacy makes them an indicator species, among the first to react to environmental change. And yet, the most recent thorough study of the bay scallop was the original and definitive Belding report of 1910 to the state of Massachusetts. But the most disconcerting piece of information is that last season was the worst in recorded history, with 6,800 bushels landed, the bottom of a steady ten year decline. By contrast, in 1980-81, 117,000 bushels were landed. Most scallop fisheries in the Northeast are closed, and many may never reopen. Why the decline here on Nantucket? Theories abound. The current political hot-button issue is nitrogen runoff from fertilizers. Many aquatic stocks, Nantucket scallops included, are naturally cyclical. I learn about dwindling eel grass, visible growth rings and nubs, fluctuating salinity levels, eutrophication, genetic diversity, the gonad index, seed distribution and propagation, biomass, the effect of storms, predators, brown algae, surfactants, oil and grease, road runoff, phosphates, overfishing. I learn that Madaket Harbor was closed to scalloping last year because it was all 'seed', and that in Dave's estimation, last season 'every adult animal came out of Nantucket Harbor.' And, in five days, when the season opens, we will find out if this will be the Nantucket scallop fleet's last season.

I ask Dave if I can go out with him on his rounds on Monday, opening day.

"Well. Any day but Monday. I'm going to have seventy boats out, and it's the first day, and the guys from the state are going to be here..." Dave's thought trails off, and we arrange instead to meet Tuesday, but I ask if I can check in periodically on Monday to see how the day is going.

"Sure, if you can find us."

The view from Dave Fronzuto's office at the town pier. He patrols the harbor in a white Sea-Ox with a short, stubby bow and a split windshield held together with duct tape. A decal with two crossed anchors and a lighthouse says: 'Harbormaster. Semper Vigilans.' Always vigilant.

November 1

Sunrise: 6:11   Sunset: 4:36
Temps: Low 43°   High 62°
Conditions @ 7:53 am: 57°, clear, wind N 10mph
Marine Forecast: Wind N 10 kts becoming SE 10-15 kts this afternoon. Seas: 1'-3'

dock patrol

I stand at the end of town pier at dawn, thumb on the red button of a brand-new, hand-held, Sony tape recorder, poised to chronicle opening day of Nantucket's commercial scalloping season. I wait, imagining and anticipating some sort of ceremonial launching of the largest scallop fleet in the Northeast. Rob is somewhere on, or perhaps even above, the island, hoping to capture this veritable armada of scallop boats fanning out over a silvery ocean. Alone at the end of the pier, these are the first words I actually record:

"Well ... except for me and one seagull and two scallopers, it's not exactly the beehive of activity I was imagining - some formation heading out into the harbor, their wakes trailing behind them like a flock of geese going off to harvest the scallops. It's really just kind of two guys. Little bit of a red glow in the east, sun will be up any minute now. Seagull's got himself a scallop, by the way."

It is my first understanding of how easy it is to romanticize the sea. Rather, I begin to appreciate the work of scalloping as precisely that - work. This is the morning commute; nothing more, nothing less. So it is from this newfound perspective that I study these two lone scallopers; their clothing, the distinctive 'A-frames' that rise up from their smallish open boats, and their various ropes, winches, nets, baskets, and other miscellaneous gear which are so picturesque in photographs, and which I now begin to see have real purpose.

As I look north across about a hundred yards of still water, an old Volkswagen van crunches to a stop on a gravel and grass parking lot called Petrel Landing. A third scalloper emerges and walks down to a floating platform, where he is picked up by a friend in a Boston Whaler sporting a Mercury 40 horsepower engine. With little ceremony, they chug out into the harbor. This boat looks surprisingly frail, and the engine barely seems able to push it.

At one of town pier's several finger-like docks, a man wearing brown waders kicks his Mercury 90 engine over. This old scalloper's A-frame looks like an ancient, weathered cross. He heads east exactly at dawn, as though delivering a crucifix to the rising sun.

In all, only four boats travel out into Nantucket Harbor - the two that I have just watched, and two distant ones that angled in from either the Boat Basin or Children's Beach beyond that. First sun now grazes the point of the Congregational Church steeple and warms the row of white clapboard mansions on Orange Street, built on a bluff there by whaleship captains to view the harbor and their vessels of fortune. The gold-domed Unitarian Church clocktower says six-twenty.

The annals of fishing tell of a cold, dangerous occupation pursued by hard, solitary men. They seldom ask whether the sheer beauty of the daily commute touches a fisherman's soul. Standing at this dock in the early morning light, I fail to see how it cannot.

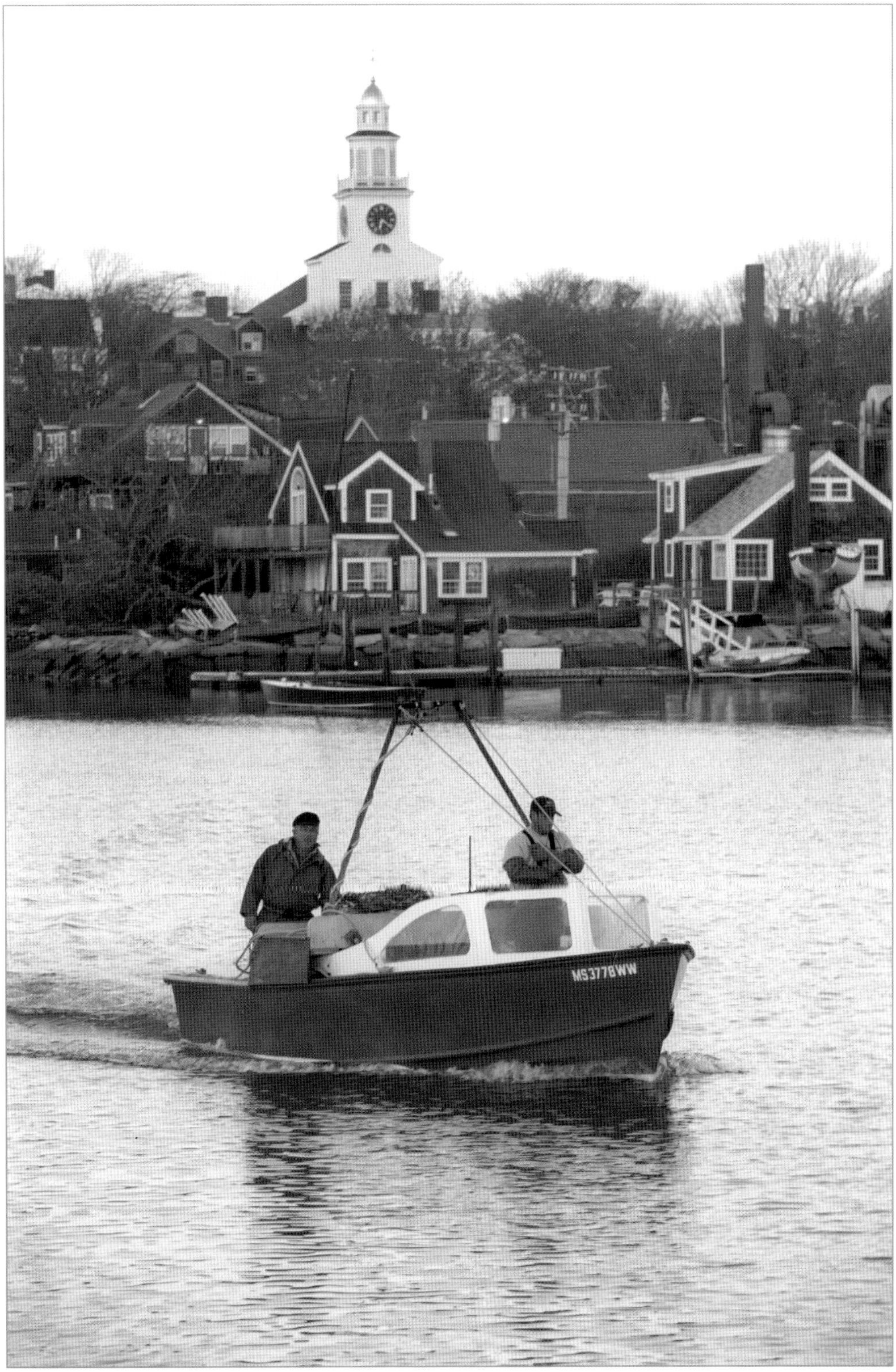
MS3778WW

The Marine Department at the foot of town pier is not yet open. A telescope is visible through the French doors which open out from Dave Fronzuto's second floor office onto a breathtaking easterly view down the full length of the harbor. A coffee-stained sign on the ground floor entrance says ON THE HARBOR, with the handwritten words *or shellfish patrol* squeezed in. A weather report, posted daily, promises 60 to 65 degrees today, extraordinarily warm for this time of year.

Town pier stretches long and thin into unprotected water. It is very much a working class pier, a little far from the action for summer yachtsmen. They prefer to dock at the privately owned Nantucket Boat Basin, a stoutly bulkheaded, expansive, rectangular enclosure that provides summer shelter to two hundred or so boats, ranging from modest family cabin cruisers to corporate yachts that positively mock the descriptor 'extravagant'. Some even come with their own helicopters, and have been known to transport political candidates, well-heeled enough to finance their own campaigns, to and from their modern day whistle-stops.

A sawtooth line of shingled cottages fringes the Boat Basin. Two cottages among this ragged line are named *Scallop* and *Cranberry*. *Scallop*, the smaller of the two, was used near the turn of the century as a studio by a woman named Elizabeth Saltonstall, founder of the Nantucket Artist's Association. It is slated to be torn down in February, but will be given free to anyone who will take it away by then.

Beyond Petrel Landing are Harbor Fuel's storage tanks, which some feel will eventually blow up the Boat Basin, and with it most of nearby downtown. Having already burned down once in the Great Fire of 1846, Nantucket has no desire to relive the incident. But, discord in town affairs being somewhat of an American birthright, nobody can agree on where else to put the tanks.

A rich tradition of local civil discord actually dates back nearly to Nantucket's settlement in 1659 - specifically to the Half-Share Revolt of the 1670's, which landed Peter Folger, Benjamin Franklin's grandfather, in the hogpen: also the jail in those days. Ten men known as the proprietors, led by Tristram Coffin and Thomas Macy, had purchased Nantucket from Thomas Mayhew of Martha's Vineyard, who had received it as a land grant from Lord Sterling. But neither Lord Sterling, nor Mayhew, nor the proprietors actually owned Nantucket. Rather, it still belonged to the Wampanoag tribes who lived there, and they didn't believe in land ownership. What the proprietors bought from Mayhew, for thirty pounds and two beaver hats, was the right to negotiate with the natives. To this end, they needed Peter Folger to translate and act as clerk, for which he received a half-share of land and a half-vote in governing the new settlement. As other so-called 'half-sharemen' were needed, eventually numbering fourteen, their influence grew. But Tristram Coffin and his two sons, Peter and James, effectively controlled the island through the proxy votes of many proprietors who never settled. The chief sticking point was that the half-sharemen could not buy land from the natives and thus increase their opportunities, as the proprietors could. Class-based hostilities emerged, ranging from a taboo against marriage between children of the upper-class proprietors and the working-class half-sharemen (despite the obviously limited choices on an island with only a handful of families), to suspicions that the proprietors were offering rum to the natives to entice them into bad deals and thus steal their land. Peter Folger's crime was to confiscate and hide the town's deed book, planning to show it to the King's representative, Governor Lovelace of New York, as evidence.

The half-sharemen, represented by fisherman Richard Gardner, eventually won the right to full representation. To add insult to injury, the Governor decreed that Nantucket's settlement be called Sherburne, after brothers Richard and John Gardner's place of birth, a slap in the face to Tristram Coffin and his sons. Resentment ran deep for many years. The first deed book was lost, and has never been recovered. Neither, some might add, has civility in town affairs.

the fastest man on the water

My own commute takes me about five miles through town and across undulating former sheep commons to Madaket Harbor, on the western end of the island. The Walter S. Barrett Pier is here at the end of F Street, near Chapel Way, pronounced as in maple, because it is named after the Chapel family rather than a place of worship.

Barrett Pier is a short asphalt wharf with docking space on two sides, and a trailer ramp on the third. Four or five boats come in and offload. Offloading is some variation of the following procedure: a scalloper pulls his boat expertly up to the dock, retrieves a parked truck and backs it onto the pier, hoists his five boxes of scallops up onto the pier, lifts them into the truckbed, moves the truck off the pier, takes his boat out to its mooring, secures it, rows back in the dinghy that has been left at the mooring since

Carl Sjolund is the first scalloper I meet in Madaket on opening day. It is no coincidence - he is the first one in on most days. The wire baskets are used to throw the keepers into when culling. The bushel boxes full of scallops are offloaded and taken to a shucking shanty for opening.

departure, and leaves the dinghy on the muddy shore. People working in pairs usually have the second person drive the truck and lift the scallops into it. A few scallopers back a trailer down the ramp into the water and winch their boat onto it, taking it home every day with their limit of scallops still in the boat. One affable man accomplishes his offloading and mooring with startling efficiency. His name is Carl Sjolund, pronounced Sher-land. Then a scalloper roughly my age says he knows me.

"You remember that tennis court you fixed up out on Dukes Road? The big red house? I was living in the house." Conversations like these are typical in a small island community. And the fact that it happened at least fifteen years earlier serves to remind that virtually everything you do will sooner or later be remembered by someone. Which tends to keep you honest.

This scalloper's name is Bruce Cowan, and when I ask him what his next step is today, he tells me he is going to go shuck, or open, his scallops. He shares a shucking shanty with Carl Sjolund, and invites me to meet him there around noon. Bruce finishes loading his bushel boxes into his beige Toyota truck, into which he has built a rectangular plywood box to protect the bed. Most scallopers have such a box. Bruce also offers to take me scalloping with him on Friday.

Bruce Cowan is happy with his five box 'legal limit', which he caught in just two tows. Fishermen say they 'catch' scallops, as though the shellfish could somehow get away. It seems a token of respect.

As I leave the dock, three breathless young men from a CBS news crew rush up. Media have swarmed the island to cover last night's Egyptair crash sixty miles offshore. But the search for debris has shifted to Point Judith, Rhode Island, and they are now looking for other stories.

"Where are the scallopers working?" they ask. They look annoyed when told they'll have to find a boat and go out on the water. But if they are willing to settle for a scalloper offloading his catch, I point them to one I've just met named Marty, who bears a striking resemblance to a young Robert Redford. I have already discovered, though, that Marty is a lot like the solitary characters Redford often plays. They won't get any sound bites out of him.

growth ring

I return to town pier at about 10:00 a.m. Dave Fronzuto and Tracy Curley are in the office with Mike Hickey, chief shellfish biologist for the Massachusetts Division of Marine Fisheries. He is about fifty-five years old with a deep and mellifluous voice, the kind of voice people imbue with authority and trust; the kind of voice incapable of telling a lie. Yet within fifteen minutes he will unashamedly tell Dave Fronzuto a whopper.

Mike has come to Nantucket on opening day because there has been a three-year running dispute between the scallopers and Dave Fronzuto's office, and even among the scallopers themselves, as to whether certain Nantucket scallops are adults or large seed. It is commonly understood that a scallop has a two year life cycle and can only procreate, or 'spat', once, during its second summer. The 'growth ring' forms when a scallop stops growing during the winter. Hence, a scallop with a growth ring has already spat, and is an adult. A scallop without a growth ring is called a 'seed'. No one seems to disagree that a legitimate 'adult' should be caught, because it will not reproduce and might even compete for resources with the remaining seed. The dispute comes from whether a large scallop might still be an adult, even if it possesses no visible growth ring, or only a tiny growth ring near the base known as a 'nub', a term in use only on Nantucket. The state, represented by Mike Hickey, has been asked to step in.

"Where is a growth ring supposed to be?" I ask.

"If you have three-inch scallop, 4/10 of an inch from the edge to 2 1/2 inches." He shows me three scallop shells on the Marine Department's brochure shelf. One is a large shell with a growth ring, one an equally large shell without a growth ring, and one a smaller shell with a growth ring.

"And if it's a problem?"

"Some years it's been way down a little over 1/8 of an inch."

"Do you go to every other harbor?"

"No, we don't normally have the same problems."

We are joined by Neil Churchill, an affable, tall, dark-haired man, probably in his forties. Neil is also a biologist for the Division of Marine Fisheries, who has responsibility for the South Shore area, Barnstable, and Nantucket. He visits Nantucket five to eight times a year to sample water quality, conduct sanitary surveys, and exchange data with Tracy. We are soon walking down the pier toward the man with the wooden cross rig, who has just docked.

"A lot out there?" asks Mike.

"I couldn't tell you. Busted my donkey engine." The man is well into his sixties, maybe even seventies, and has a soft gravelly voice towards the high register. He is clearly disgusted, and wary of three strangers looking down into his boat. I

Any scallop without ...
"...a well-defined raised annual growth line...shall be deemed seed."
Massachusetts Statute Chap. 130, Sec. 70.

wonder what a donkey engine is.

"I'm sorry to hear that," says Mike.

"Not as sorry as I am, mister."

"Whereabouts were you fishing?"

The man mumbles an answer, and refuses Neil's offer to help lift his box up onto the shoulder-high dock, although a full bushel box looks like it weighs a good fifty or sixty pounds and a helping hand might be welcome.

"If it drops, it's my fault," the man says. "Are you a warden?"

"No, I'm a biologist with the Division of Marine Fisheries."

The man's tone immediately changes.

"Oh, okay! Take whatever you want. You may find some seed in there - I don't know."

The old scalloper is now willing to divulge that he was fishing off Hussey Shoal, as long as we don't tell anyone else. And he wonders if they could tell him what the small, crablike creatures crawling on his catch are. The state biologists spend about ten minutes measuring scallops, and marveling at how big and deep this year's crop seems to be, and how well-defined the growth rings are. The average length of the shell is about seventy millimeters from the hinge, with growth rings averaging fifty-five millimeters from the base.

"This is the best stuff I've ever seen," says Mike flatly. I figure in twenty-some odd years, he's seen a lot of scallops.

Neil finally comes across an adult-sized seed, the only one out of perhaps a hundred they have now checked. With a knowing look, he hands it to Mike, adding mysteriously, "I thought you might want that." Mike grins and slips it into his pocket.

We head back down the pier toward where Fronzuto, though engaged in animated conversation with a boater, clearly has one eye on us. Mike takes the large seed out of his pocket and hands it to him.

"You've got a whole harbor full of this stuff."

Dave glumly takes the scallop, chuckles in resignation, shakes his head.

"Yeah. Big seed. Big, big seed." He is imagining an unpleasant *deja vu.* Last opening day, following what was, until then at least, the worst scalloping year in history, Dave, his wardens, and the state officials were literally surrounded on the dock by a hostile mob of scallopers. Although there were a number of scallops large enough to be adults, the growth rings were all but invisible, meaning most of the catches would have to be thrown back if the law was rigidly enforced. Against their better judgement, it was determined that discretionary slack would be given regarding the so-called 'nub' scallops in the hopes that a more normal growth ring would appear this year. But the shell in Dave's hand is clear evidence that it has not.

"About two-thirds of what he had was that," says Mike solemnly.

"Jesus Christ." Dave's eyes are fixed on the seed scallop in his hand. Mike can't keep it in any longer.

"I'm a liar," he finally confesses. No matter, he has played his hand perfectly. Hard laughter erupts, Dave's the loudest of all.

"Oh, you bugger! You rotten bugger!" Mike now shows Dave a typical scallop. "Oh, that's beautiful," is the reaction, one of admiration as much as relief. "That's beautiful."

"That's average," brags Mike, as though he grew it himself. "One, out of...how many did we measure - a hundred?"

"That's beautiful. Oh, that's beautiful." Dave repeats it, soothed by the sound, and then turns to me. "Is the tape recorder on? Is it on now?" He leans in to the chest pocket where I keep it. "You rotten son of a..." he says as though Mike were in the recorder, then to me. "If you realized the grief that we've gone through over this issue, for years, then you'll realize why I called him a rotten son of a... How many boxes did we look at for the past two years that were all like that? We fought this battle, and we didn't do anything good for the fishery by letting them take that stuff - we really didn't." Dave looks down at the typical scallop, which he is still carrying as though it is a piece of ribbed treasure. "Oh, that's beautiful. I am so happy. He got me bad, too."

"Is the tape recorder on? Is it on now?"

He leans in to the chest pocket where I keep it. "You rotten son of a ..."

Frank Marks and George Rezendes

the four million dollar man

At a floating dock below us we see two more old timers: lean George Rezendes and sturdy Frank Marks. Dave whispers that he was George's assistant Harbormaster for years, and then jumps down, overenthusiastically extending his hand in a faked greeting.

"Hi, I'm Dave Fronzuto! Damn glad to meet you!"

George responds in kind. "Hi David, how the hell are you?! You come here often?"

"Once in a while," laughs Dave. "Frank? Three quarters of a box!?"

"You got damaged goods here you know."

"Oh, in a pig's ass. You been rebuilt, Frank. They've completely reshimmed you. A four million dollar man!"

"No. But I bet a million and a half. I been there three times."

This cryptic conversation intrigues me, but I am more eager to pursue the possibility of my first interview with a scalloper. In October I had tried to interview two or three whom I had seen fixing up their boats. They had deflected my questions, sometimes rudely, claiming there wasn't much to tell, or referring me to the true old-time scallopers like Eddie Stojak. I am therefore anxious not to blow this opportunity, but Frank has already gathered up his gear. I make a point of refreshing my memory that his name is indeed Frank. Nonetheless, in what would prove to become an unfortunate hallmark of my interviewing style:

"Fred Marks?"

"Frank."

"Sorry - Frank. I'm, uh, hoping to write a book called *Scallop Season.*"

"Uh-huh."

"Essentially just a diary, chronicle of what goes on during the course of the year."

"My son-in-law, he's editor of that *Nantucket* magazine."

"Your son?"

"Son-in-law."

"Son-in-law?"

"Shep."

"Shep is your son-in-law?"

"He wanted some guy to interview me too. Told him I didn't really have the time." Frank chuckles at this memory.

I am discovering that the printed word fails to convey the Nantucket scalloper's eloquence. Speech rhythms, varying pitches, comic pauses, emphases on crucial words, speeding up here, slowing down there, dragging one syllable into two where needed, adding a question mark to a statement, an exclamation point to a question, or a chuckle to a sad tale: these tell the real story.

"We're hoping to go out, you know, several times this season, take pictures? Maybe we could go out on a boat with you ..."

"I tell you who might have a pile of pictures. I don't know what he did with 'em - was, uhh - Stan Grossfeld. He went and shot a lot of pictures. I don't know who the hell had 'em at the time. But shot a shitload of them."

Frank is already walking down the dock toward the parking lot. Grossfeld, now a Pulitzer Prize winning photographer for *The Boston Globe,* had published one of my favorite photo books called *Nantucket, The Other Season*, in the late 1970's, and I figure this must be what Frank is referring to.

"You want me to carry something for you?"

"No, no, I got it." Frank has moved ahead so quickly that I now have to walk briskly to catch up.

"So you've been scalloping how long?"

"Since '69."

"Do you go all season long?"

"I used to. I've had three heart attacks. I just had open heart surgery in May."

"Three?! Is that what you were just talking about getting rebuilt?"

"Damaged," he chuckles.

"So you have to take it easy?"

"Well I'm *supposed* to take it easy."

Frank still runs a small landscaping and caretaking business, and George did all his grasswork while he recovered. I ask him about today's scalloping.

"Just a sport. But I'll tell you one thing. That scalloping put me on my feet. My wife and I worked every day together twelve years, 'til the taxman told her one day she's working fifty cents on the dollar. She never went back in the boat from that day to this. That's quite a kick in the ass, you know."

"Did you enjoy scalloping with her at the time?"

"Yeah," he says wistfully. "Not only that, I took my brothers, I took my neighbor. I kept going."

"Was that something you had to pay all at once?"

"Oh, you gotta pay."

Frank Marks

David Coombs

donkey engine

Frank notices the man with the cross-shaped rig, whose scallops we first checked, and tells me that his name is David Coombs. He is pushing a wheelbarrow with something that looks like a lawnmower engine in it. I gather that this is a device called a donkey engine, and that it is critical to the scalloping operation, since having it break down has ended David's fishing day, but I do not know what exactly it does. At the same moment, David notices Frank coming down the dock and calls ahead.

"Oh, so you're the one that was out with George! Where's your boat?"

"Over in the yard." Frank repeats it sadly. "Over in the yard."

"You don't have a donkey you want to sell, have you?"

"No. What's the matter with this one?"

"She ain't got no compression. She got nothin'."

"Take the spark plug out and throw it overboard," deadpans Frank.

"Huh?"

"Take the spark plug out and throw it overboard."

"No compression," repeats David mournfully. Frank realizes David isn't going to get the joke, and changes his tack.

"Well, the rings are bad if you got no compression. Or you got a stuck valve."

"It just went all of a sudden. I finished hauling back and I went to start it again and nothin'. Spun right over, just slick as grease." David tries to start it a couple of times to no avail. Frank nods.

"My last tow I pulled by hand - from the deepest part of the harbor!" says David. "Ho-ho - Jesus! I don't mind gettin' 'em up to the boat - it's gettin' 'em from the rail to the cullin' board!"

"Yeah, that gets to be heavy there," Frank agrees as though from experience.

"'Course I'm one of the new boys, but I understand now why we use such long shanks. It's so we can fend 'em over the rail and drain 'em."

"That's what we did. I held my knee on it."

"Thank God, you know I only go six rings deep. I'm glad I don't go seven rings deep."

"Well, we had two small dredges that fished very good. But we had two big wide dredges that didn't fish for shit. But we don't have no weights on 'em."

"I was catching most of my scallops in the first and second dredges. The back dredges weren't catchin'."

"I told George 'I got plenty of dredges, you need dredges'. He said four's enough. He's right."

"Yeah - oh yeah. Pain in the ass, especially if you're handhauling. Course, years ago there wasn't any god damn grass around here. These guys used to sail these boats, these old catboats, and they'd tow all the the way down the harbor and back again, and they'd haul back all by hand. But it was all shell. They had long A-frames on the front of the dredges instead of the short ones they got now."

"How long you been scalloping?"

"About thirteen years."

"Always haul in by hand?"

"No! No!" He says this as if I'm crazy. "Use a donkey, but it's broken. That's for the birds!"

As Frank gets in his truck and drives away, I realize that he has successfully avoided talking about himself. But at least I now have some idea what a donkey engine is, and that it is used to pull the dredges back into the boat after they have dragged along the bottom and filled with scallops. But this is the first time I have heard most of these other terms, terms like 'fend, handhaul, haul back, shanks, culling board, rail, rings, tow'. I have already arranged with two scallopers to go out with them on Wednesday and Friday. Maybe I will learn a little more then.

Dave Coombs regards his compressionless donkey engine. "Oh, dear," he says, mostly to himself.

Rings are linked together to form the basket of a dredge. David Coombs' dredges are six rings deep.

MS 12 54 KM

Eel grass piles up everywhere along the shore, blown in by winds and tides. Until this week, it had just been seaweed to me. But I have since learned that it is not only the preferred habitat for scallops, it is useful to humans as well. In winter, people mound it around plants, foundations, and fire hydrants, like nature's insulation.

eel grass

Returning to the Shellfish and Marine Department office at the foot of the pier, I find Tracy Curley, Mike Hickey, and Neil Churchill. The scalloper with the Volkswagen van has returned and dropped his partner at Petrel Landing, and we embark on about a hundred yard trek to go check and measure his catch.

Nantucket is one of the few towns in the state to have its own biologist. Tracy, a Nantucket native, went to Wheaton College and then was trained by the National Marine Fisheries Service for the Foreign Fisheries Observer Program. She did a stint whalewatching in Maui, has been a fishery observer, has scalloped, and takes her shifts every season as a shellfish warden. She sees no gray area in the growth ring debate. 'If you have to look for it, throw it back' is her motto.

I ask Tracy how her day is going, not as protective biologist, but as shellfish warden on the first day of scallop season.

"Right. So it's very stressful," she laughs. Tracy has a direct and forthright, though not particularly loud manner. I get the feeling she would prefer to direct attention away from herself, but if a situation arose that demanded she attract it, she would. Tracy tells me that her first order of business had been paperwork: filing a Notice of Intent with the Conservation Commission, the state-regulated wetlands protection body, to allow the Marine Department to brushcut a parking lot for a pier being rebuilt at Jackson Point in Madaket. She then went to Children's Beach, her usual area, and checked one scalloper's catch for seed.

We walk across a wet and spongy carpet of flat, brown, stringy seaweed, which lays here several inches deep and several feet across, and which I have noticed for as long as I have lived on Nantucket. It rings the island, piled up wherever the sea meets the land, stacked and jumbled up near the top of every beach, sometimes two or three feet deep, a good marker of the highest tide. It has the look and consistency of brown confetti, and is particularly deep where a storm has blown it ashore, as though the sea creatures threw a giant party and forgot to clean up. I have also heard the term 'eel grass' as long as I have lived here. I listen to a local bluegrass band named Eel Grass. Only today do I put two and two together, asking of Tracy, the marine biologist, what must seem a horrendously stupid question - whether what we are walking on is, in fact, this ubiquitous, legendary eel grass. She replies with the courtesy of a scientist that indeed it is.

Eel grass, as it happens, is the scallop's city. It lays along the bottom in shallow water, swaying with the tidal currents. And it serves two extremely important functions. It provides a home for baby scallops, before they have fully developed shell structures or size. With tiny ligaments called byssal threads, the almost microscopic scallops attach to the eel grass, which keeps them from getting swept away at the whim of every tidal current or storm. The eel grass also slows down the currents, which allows the maturing scallops to filter more food out of the water. Most scallops, even once they have grown enough to detach themselves and forage for a new location, will still opt to remain in the eel grass. As it turns out, this will not be the last time I hear the term.

So ingrained is it in the island sensibility, there is even a bluegrass band called Eel Grass.

Tracy Curley

if you have to look

When we are close enough to the scalloper whose catch we are trying to measure, I recognize him as a younger guy, maybe thirty, whom I associate with partying and surfing, though I have never actually seen him at a party or with a surfboard. I did, however, once help him right his jeep after he had flipped it doing doughnuts in the Jetties Beach parking lot. He yells that he is going to unload at Grey Lady Marine, an industrial looking boat maintenance facility on the other side of town pier. We are forced to retrace our steps, and by the time we reach him, the van is waiting and two boxes have already been loaded in. When Tracy tells him that the two state biologists would like to measure his scallops, he looks both nervous and ticked off. He is completely disinterested at my first attempt to find out how his day went. However, a fuel tanker arrives and blocks us in while noisily delivering thousands of gallons of fuel. Thus trapped, he reluctantly agrees to talk, anonymously. I get the feeling I am not going to get much, and I am right.

The state biologists attempt some banter by asking where he was fishing, to which he gives a non-reply. Then Tracy points out that his bushel boxes are cracked and warped, thus holding too many scallops, and tells him that he'll need new ones. He uses this opening to spring the $64,000 question, with well-practiced sarcasm.

"What are we calling a well-defined growth ring this year?"

Tracy instantly picks two scallops off the top of his box. "This is well-defined, this is not."

"That's not a well-defined growth ring?"

"That's not a well-defined growth ring."

"I can see that coming out of the water."

"That is not well-defined." Tracy indicates the clearly marked good scallop. "This stuff is obvious," she says. "If you have to look, don't take it. Okay?"

"You really should notify people."

"All the meetings are public."

"I'm not trying to make an argument here. All I'm saying is there's only so many commercial licenses. The mailing address is right there when you buy your license."

"You want us to mail out two hundred notices about well-defined growth rings, which is what you should be taking anyway?"

"I know it sounds like a pain in the ass for you guys to spend a day in the office doing something like that." His tone now softens. "I'm not trying to be a hardass."

This argument has been peppered with laughter by the listeners, and it is clear it hasn't been the first for either. Mike and Neil have measured their scallops, the fuel truck has moved, and we are on our way. I ask Tracy if the growth ring regulation hasn't been the same for a number of years.

"Yeah, but we have to go through it every year. I mean, he did have a point, but..." Her voice trails off in resignation. Fishermen do not want to be regulated, and her job as a warden requires regulating them. And her position as town biologist makes her more aware than anyone on the island, if not anywhere, of the danger that the Nantucket Bay Scallop is in from a variety of sources. Given the option of continuing to regulate the fishermen or looking the other way, the choice in her mind, like a well-defined growth ring, is obvious.

Tracy Curley checking growth rings - only five per cent may be 'seed'. Most fishermen have learned that old, broken boxes hold more scallops.

Carl Sjolund's shucking shanty

shucking shanty

For years Nantucket had two banks, Nantucket Bank and Pacific National. One might wonder why an island so intrinsically identified with the Atlantic Ocean would have Pacific National as one of its major banks. Very simply, it was founded with money brought home from the vastly lucrative Pacific whaling trade, discovered and dominated by Nantucketers. The term 'Plum Puddingers' referred to Nantucket whalemen who had not yet rounded Cape Horn into the lush Pacific, perhaps because their wives liked keeping them closer to home. Indeed the Pacific had its temptations; the charming and exotic women of the South Sea islands captivated many a young whaler. The name Cape Horn was, by some accounts, derived from an axiom that Nantucket men hung their morals upon it like a hat on their way out, and picked them back up on their way home. A similar instinct led the *Bounty* mutineers to settle on Pitcairn Island, where they were discovered years later by a Nantucketer named Mayhew Folger.

As danger goes, the Pacific also held its share. In the famous saga of the whaleship *Essex*, inspiration for Herman Melville's epic *Moby Dick*, the suddenly shipless crew chose to row their open boats an impossibly long distance against prevailing winds and currents to South America, rather than the much more logical direction toward the Marquesas Islands, because they feared encountering cannibals there. Ironically, as retold chillingly in Nantucket historian Nathaniel Philbrick's account of the incident, *In the Heart of the Sea*, some of the men themselves resorted to cannibalism to forestall starvation anyway.

All of which serve to explain not only why there is a Pacific National Bank at one end of Main Street, but also a Pacific Club building at the other, a seafaring men's social club which until a year or two ago still met in the downstairs room. To show the high regard in which the Pacific name is held, the bank's moniker has survived intact two recent takeovers by two successive New England financial behemoths - testament to its tenacity, like that of the whalers who founded it.

Standing in line at the same Pacific National Bank last week, I saw a man wearing a T-shirt that said:

Give a man a fish and you feed him for a day.
Teach him to fish and he will sit in a boat and drink beer all day.
Fish 24:7

The man's shirt implies that fishing is a leisurely occupation. A scallop fisherman's day is somewhat different than that of a summer sportfisherman's outing. Most scallopers are in their boats before sunrise, docked and offloaded by 9:00 a.m., and beginning the next step in the process: opening, or 'shucking', the scallops. Some scallopers open their own, but a whole cottage industry of shuckers operates from dozens of small shanties tucked away behind homes, beside garages, or in any other nook big enough to hold one or two people and a pile of scallops.

Carl Sjolund has such a shucking shanty, nestled among pines behind a house he built and has named 'Homeport'. It is a small shed, roughly eight feet by ten, very clean and white inside, with a faucet and a refrigerator, a drain in the concrete floor, and a fiberglass counter the whole length of one side. The right side of the counter is already piled high with Carl's scallops, who has dumped them off and then run some errands. I meet Bruce Cowan here and help him bring in and dump his boxes, building an equally high pile on the left. Bruce picks up some of Carl's scallops, and his generally unexcitable voice rises in amazement.

Bruce Cowan

"Look at these! Beautiful! Look at these things! Oh my God - these are nice!" Bruce explains that Carl's scallops are much bigger than his, but notes that the best scallop is actually small but deep, so more can fit into a box.

Entering a scallop shanty, one is immersed in a thick, languorous smell, rich and textured with ocean salt, virtually liquid with the unmistakable sweetness of a scallop. Then, one becomes eerily aware of a light, almost jazzlike tapping. Scallops are constantly opening and snapping shut. Several thousand create a ragged, yet pleasing, percussion. Finally, one feels watched. If you look closely at an open scallop, in the ridges just inside the shell's edge, you will see a whole battery of bright blue eyes peering out. The scallop is examining you as intently as you are examining it.

It is legendarily difficult to open scallops. Bruce tells me that if you are right-handed, you put the dirty side of the scallop (the side that rests up in the water, where moss and algae grow on it) up, and insert your knife into an opening near the hinge. The adductor muscle, which is called the eye, and is the part that is eaten, is closest to the right side in all scallops. You then angle the knife up to press against the top of the shell, and sever the eye cleanly against the shell. To remove the gut, you very lightly and carefully insert the knife on the back side of the eye, using your thumb to hold the gut against the blade of the knife, and pull it towards you. If you do it well, you can get all the gut in one stroke, and drop it in the trash barrel at your side. You then sever the bottom of the eye, and flip it into bowl in front of you, dropping the shell into the barrel also. If you are deft, the operation requires three strokes and you get a clean, whole eye. If you are not, you take about twelve strokes and get a torn scallop with some guts mixed in. Multiply that by several thousand and you have a long day shucking ahead of you.

It is legendarily difficult to shuck scallops.

five decades of scalloping

Jim and Eric Worth come into the shanty. Jim is Carl's cousin and Eric is Jim's son. Jim is somewhat quieter than either Bruce or Carl, and Eric takes after his father. Whereas Bruce had just opened a couple of scallops to demonstrate the technique for me, Jim is clearly here for the real thing. He wears oilers and boots, and dons latex gloves. He puts a trash can by his feet on his left side, a mixing bowl type of container on the counter, and shovels a pile of shells in front of him. I ask if he minds my watching him work and talking to him.

"If it starts breaking my concentration, I'll tell you to knock it off. I haven't opened in a few years." I ask him how long his family has been here and scalloping.

"My family goes back to the original settlers on Nantucket. First marriage on the island was a Worth and a Macy, I think. Had a lot of whaling captains and everything else on my side of the family."

Given what I've heard about shucking, what I'm now seeing is unbelievable. It takes Jim about two seconds to open a scallop. Bruce had demonstrated in slow motion, and I saw the procedure clearly. At full speed, it looks like the work of a magician or a card shark.

"A lot of these guys go a lot faster than this - they've got something in the air all the time," says Jim.

When Jim says something in the air all the time, he means that the shell goes out of his hand, the eye goes flying into the bowl, and seemingly before either hits the growing pile of shucked scallops or empty shells, he has another in his hand. Jim no longer scallops, but used to fish with his father until the early 1970's, on Saturdays or over Christmas breaks as a teenager, and during furloughs when he was in the service. As if on cue, a friend pops his head into the open doorway and announces that he has some old black-and-white photographs of Jim's father scalloping. Jim points to one.

"That's Dad in his dory right there. That's when scallops were scallops." He explains that the yield per bushel was two pounds per bag more in those days; and that they used burlap bags instead of bushel boxes. Scallops are measured by the number per pound, or count as it is called. Average Nantucket scallops come in at about eighty per pound, or 80-count. So-called 'beans' might come in at 120-count. Jim tells me they used to get 58-count scallops. An average bushel box nowadays seems to net around eight pounds of shucked scallops.

I have begun to gather that Jim's son Eric is here to learn. Eric is twenty-two years old, and although he grew up in a scalloping family of many generations, he hasn't done it yet. Bruce needles him a little.

"Are you gonna open?"

"I'm not that good at it yet..." says Eric, a little nervous, as though that sentence might have ended "...and I don't feel like practicing in front of a stranger, either."

"He's just learning," says his dad, his eyes twinkling. "He didn't bring his boots with him. But, he's got his truck here. Here's the keys if you want to come back. Don't be shy." Everybody gets a good laugh at this. Eric does, in fact, stay to open.

Relentlessly cheerful Carl Sjolund now fairly bursts into his shanty, ready to work. He is around fifty, but looks and acts considerably younger. He has Norwegian features; blonde hair going white in places, ice-blue eyes, a strong, square open face. I have just read that, on Cape Cod, fishermen of Norwegian descent are called 'squareheads'. I figure he'd laugh at this. Carl has a bumper sticker that says: 'I can only please one person per day. Today is not your day. Tomorrow doesn't look good either.' Ironic, since Carl is also a caretaker - which means he watches over a number of million-dollar properties that are unoccupied most of the year, and thus subject to a variety of winter insults, ranging from vandals to nor'easters, birds down the chimney to bugs on the carpet, sticking doors to bursting pipes. Caretaking carries a lot of responsibility, and requires a vast storehouse of skills, knowledge, and connections to be good at. Almost no one is, and as a result homeowners are forever begging for recommendations of good caretakers, and in desperation plead with friends, relatives, whoever they can convince to watch over their houses. But I suspect Carl is not only very good at caretaking, but very pleasant and accommodating to his clients. They probably chuckle at the bumper sticker also.

Carl has been scalloping for forty-two years and is one of the names everybody mentions when they think of legendary scallopers. This is Bruce Cowan's paean to Carl's fishing skill: 'He's unbelievable. I fished right next to him, the exact same spot. Taking the exact same tow, side-by-side. I'd practically have a coronary trying to keep up with him.'

Carl is just as efficient with language as he is with his dredges. He typically omits unnecessary words, showing particular disdain for useless pronouns or modifiers. When I ask him to name some of his peers, he is a little reluctant. But he makes a point of telling me to talk to a man named George Andrews.

Carl Sjolund taking a box of scallops from his nephew Eric Worth, who is learning the trade this season. Rope handles make the boxes easier to lift.

selling thc scallops

As Bruce, Jim, Eric and Carl are shucking, I ask the following question: "So, how does one actually go sell scallops at the end of the day? Who buys, what do you do?" I feel immediately that I have waded into forbidden territory. The selling of scallops, especially on opening day, is a sensitive issue. Bruce decides he can at least give me an overview, if not specifics.

"Glidden's, they have a shanty, they'll have openers and they'll be weighing guys' catches in there. Sayle's - he's got the same sort of set-up. Souza's - they'll be buying scallops." Bruce explains that these are the three major fish markets on the island. The physical process is that the scalloper will put his catch into one or two white five-gallon tubs and take them to the buyer. The buyer will then dump them onto a scale, weigh them, and write out a slip acknowledging receipt of however many pounds are tallied. The scalloper generally receives his check the next day. The fish market will, in turn, ship most of the scallops to a major wholesaler. A name I heard earlier today, Mike Foley, is just such a wholesaler. It was rumored that he was prepared to offer thirteen dollars a pound for today's scallops.

"And then there are a few independent guys that buy scallops on the side," Bruce elaborates. "The Nantucket fishermen have finally begun to market scallops on their own. For quite a long time there were just a few big buyers in Boston and New Bedford. The laws of supply and demand weren't necessarily being followed, because they really controlled the market. It was getting very frustrating. Guys knew there weren't many scallops out there, and we were still getting lowballed. Nothing much you could do. So over the past ten years or so, some fishermen have begun to develop their own accounts, sell them on their own. The supply to the big guys in Boston diminished, so the price went up."

Carl, in fact, has a license to sell small amounts of scallops privately. But he points out that most scallopers remain loyal to one buyer throughout the season.

"Does the price vary much from day to day or buyer to buyer?"

"The beginning of the season it can bump up a quarter, ten cents. Bump down same amount. It usually doesn't jump around too much."

"But they always all pay the same," Carl says. "One guy goes up a nickel, the other guy'll go up a nickel. And they know it, too. It's not price fixing, but by golly...."

"Word gets around. Say you have a guy that is loyal to you, and he hears that Danny so and so is buying for a quarter more. He's gonna go in and say 'How about it?'

"I've had it where the price goes up during the night, and when I came back they say 'I owe you twenty five cents a pound more for the ones you delivered last night.' But they don't take off when the price goes down. And I've never seen 'em know what the price is going to be before the season starts. It's always a big game."

"And in a way you don't really want to give your scallops to the guy who sets the price," says Bruce. "Because you want them hungry for more. They're more likely to bump the price up if they're not having the scallops come through the door. But on the other hand, you don't want to sell them to some fly-by-night person who might be offering a quarter or fifty cents more, because you don't want to get a bum check. It doesn't happen often, but you've got to be careful. Checks have bounced."

I ask Bruce where he intends to sell his scallops, and he cheerily ignores the question by telling me he will just put them in his truck and decide then. But he is insulted, indeed horrified on behalf of all Nantucket scallopers, by my next question.

"Do you freeze them?"

"No, no, no, no, no! No one does that! That's just not how it works."

"Yeah, they want 'em fresh," emphasizes Carl.

"I've heard people say these are the best in the world."

"Well, if you asked somebody on the Vineyard they wouldn't agree with you," Carl says.

"Have you had Vineyard scallops?"

"Never eaten one. Don't think I've ever seen one. But the areas that they get 'em, they've been shrinking every year. They used to get a lot of 'em on the Cape. Like Buzzard's Bay, they used to get 'em in Buzzards Bay. Marion. New Bedford. Fairhaven. Dartmouth. Used to get a lot of 'em down on Long Island Sound. Then they had the brown algae. I don't know if they're coming back. Maybe it's reached a low point because people are more aware of pollution now then ever before."

I ask Carl if scalloping could support you year-round back when he started, or if one could just quit their job temporarily during scallop season.

"And take the rest of the year off? Hah! We'd stay with it all winter. There wasn't the work that there is in Nantucket now. Carpenters would have crews - you'd want to stick with 'em because you won't have winter work. There wasn't building a house on every corner of every street like it is now. When I first started we got fifty-five cents a pound. Opening your own if you got twenty-five dollars that was good. I did carpentry work in 1969-70. I don't know if I got two bucks an hour or not. But I know I worked forty-seven hours a week and I took home ninety-nine dollars."

"So how many guys you expect to be out in January nowadays?"

"Depends if there's any ice. Depends if people have other commitments. Depends on how much they're making fishing compared to banging nails or doing what else they might want to do. Almost everybody out there scalloping now has another job. So when it drops off they can go to it."

"Lot of them self-employed?"

"You almost have to be. Otherwise you can't take that kind of time off."

"What if you get a week of real cold weather when you can't go out? It has to be above a certain temperature?"

"Twenty-eight degrees. Years ago we maybe couldn't go out for two, three, four weeks at a time. Carpenters all had their crew - didn't want somebody for a couple of weeks. You could probably do it today, though. If you knew a guy that has a painting business, you could go paint with him for a couple of weeks. So we need your phone number before you leave."

Carl grins. And he tells me again to make sure I talk to George Andrews.

During maple syrup season in Maine, men gather in sugar shacks and literally watch water evaporate. The occasional feeding of slab wood into the furnace or turning of a spigot to let a few drops of maple syrup flow out provides the major excitement. Yet, the ritual holds infinite fascination.

A shucking shanty holds a similar fascination, and I begin to understand why these long stretches of monotony appeal to men. It is so they can have conversations like the one I heard at Carl's shanty:

"Hey, what do you guys do with your shells?"

"I take mine up to Slozek's. Or the Jetties. Why? Want some in your driveway?"

"I don't need that. Not now."

"Yeah - your wife wouldn't like that. Them birds sitting on your lawn, shit all over."

"Put 'em there and then get a couple of good days of rain. Then some sun. Boy that'll smell some."

end of a long day

It is 4:00 p.m. back at town pier in the harbormaster's office. With the cold and darkness of daylight savings-shortened sunset approaching, Dave, Tracy, and the rest of the staff look exhausted. They are standing in jackets and hats, leaning against walls or sitting on counters. They would like to go home, but they cannot. There is still a scalloper on the water. I ask if it is just the one guy they are waiting for.

"It takes the same amount of effort for one guy as it does for a hundred guys," Dave growls in frustration. "I can't deal with this on the first day. Especially when everybody else in in at 9:15."

As we maintain our vigil, a lady comes in with a piece of paper that has something like a Rorschach test on it. She gets Dave to stare at four dots for thirty seconds and then look up. She asks him what he sees.

"A pumpkin." This seems appropriate, given that it is the day after Halloween. I am given the test. I see either Jesus Christ or Charles Manson. Everyone laughs, but I am genuinely curious as to which it is, Jesus Christ or Charles Manson.

"Jesus Christ, I see a pumpkin!" says Tracy.

Dave takes the test again, this time turning the paper upside down.

"Dave, it ain't gonna do you no good to turn it around," says the lady. "You see what you see. You don't see the light, you're in trouble."

"You should be able to see God everywhere," counsels Tracy.

"I raised my head and it looked like a pumpkin," Dave pleads, as though guilty of something, hoping to make amends. His thirty seconds are up, and he closes his eyes. "I see it! Yep. I see it! Just bigger than shit, I see it. Why is it getting clearer?" Near constant laughter now echoes in the small office. "It got clearer! It actually got clearer! I still see it."

"You'd better *keep* looking!"

"When I first saw it, I said - Jesus Christ! And then it like develops. It develops before your very eyes. I'll be go to hell."

"Well that's where I thought you *were* going!"

"All right, gotta go Dave," says the four-dot test lady. "Dog's gotta pee. I thought you could use that at the end of the day."

The mood is considerably lighter now, and Dave mutters more or less to me, "It shouldn't be this hard. It really shouldn't."

They trade stories about the day's events: a few routine discussions with scallopers about the growth rings; a truck that got stuck in the sand on Smith Point; a guy already giving grief to Paul Boucher, one of the wardens, about whether they're going to start cracking down about growth rings on the first day. I ask Dave if he knows what the price of scallops will be.

"Man from Foley fish told me the price was thirteen. Highest ever. But like he says to me, they're paying fourteen on the Vineyard, they're paying fourteen in Bourne, and they're not as good scallops."

"If I went over to Glidden's Seafood right now, I could find out what they actually paid?"

"I don't know if they're dickering on the price. They may not know what the price is, or they may not want to say. Might say 'I'll give you ten bucks and if it goes up, I'll make up the difference.' We'll see. He said he's going to open at thirteen. You know, he could've changed his mind while he's walking down the street."

"Thirteen certainly caught everyone's attention," I note, having heard a few people talking about it during the day.

"Oh, yeah. You figure they're landing ten boxes, that's an eight-hundred and fifty dollar day. Then you've got to pay the culler, and you've got to pay for your fuel, you've got to pay your openers..." Dave drifts off into a little more business with his wardens, then returns to me.

"What do you want to do tomorrow?"

fleet

When you are on the water, you are part of the fleet. Even if nobody likes you, you are still part of the fleet.

No particular number of boats is required to comprise a fleet. Sometimes when a storm is coming, several unrelated vessels from various fishing towns will ride it out in Nantucket, a designated 'safe harbor'. 'Fleet's in', that ancient port-o'-call herald, will be heard around the bars.

Being part of a fleet carries a special acknowledgement of inclusion. This is particularly ironic, since fishermen tend to be somewhat solitary. Or at least, the nature of the occupation forces them to be more solitary then they otherwise might be. Other trades have no such equivalent. There is no fleet of, say, plumbers, or masons.

November 2

Sunrise: 6:12 Sunset: 4:35
Temps: Low 49° High 64°
Conditions: Partly cloudy, wind ESE 10mph
Marine Forecast: Wind SE at 15 kts becoming SE 25-35 kts by afternoon. Seas: Building to 6'-12' by evening
Tonight: Storm Warning. Wind S 40-50, gusts to 70. Seas 10'-20'

harbormaster

Rob and I meet Dave on a sandy road in Madaket, by a small supply shack. Several rolls of snow fence are stacked behind it, picturesquely awaiting winter. These may be among the last of the old wood-and-wire snow fences that so lovingly decorate the eastern seaboard in winter, as they are gradually being replaced by the bright orange nylon fencing that surrounds urban construction sites. Just ahead is the summer dwelling of children's television icon, Fred Rogers. If Fred is at home today, he has a grandstand view of the scallop fleet, sixty boats with as many variations of color, design, and A-frames, departing parade-like beyond waves of beach grass, bright green blades all gently leaning away from the light northerly wind.

Several years ago, near this location, a town selectman stumbled into a controversy regarding public beach access. Most Nantucket beaches are privately owned, but most owners have customarily granted open access. And, colonial Massachusetts law grants public fish and fowl hunting rights between the high and low tide lines. Still, homeowners periodically attempt to close access to their beach property through various tactics, some legal, some questionable. The selectman's two children were evicted from this stretch of beach during one such well-publicized attempt, prompting a quote which made Boston media: 'I have the only two kids ever kicked out of Mr. Rogers' neighborhood.'

Dave drives a town-owned, bureaucrat-blue, Ford 150 pick-up, but my observations of the past few days tell me that he is far from a faceless paper-pusher. He is no-nonsense about his duties, but his gruff demeanor belies the fact that something funny is always bubbling just beneath the surface of his mustachioed face. A seldom-lit Frisco cigar is often in his teeth or his hand. Dave either oozes sarcasm or radiates genuine admiration. You are never uncertain which, you are always entertained.

"Edit out the swear words," he tells me. "I come by it honestly. Twenty years in the Coast Guard and ten years doing this. I came to Nantucket in 1984, with the Coast Guard, to run the LORAN station out in 'Sconset. And then I had the dubious honor of being involved in the automation of the LORAN station. When I came here we had a crew of thirty-four and when I left we had a crew of four. And then I went to the station that we automated everything *to*, in Seneca, New York, and I ran six LORAN stations by remote control. And then a very learned commander that I was working for said - 'Well, we did that so nice, now let's go do another dozen.' And I said - 'Well, okay, no problem. Where do you want me to go?' And he said - 'Oh, no. You're going be the staff coordinator for the whole project, for the rest of the United States.' And that was in New York City. And Dave said - 'Nope, unh-uh. Married to a Nantucket girl, so when I get done doing this, I'm going back to Nantucket.' So I did. And it was the right choice for a lot of the right reasons. My son was just starting high school, and he got to go to one high school. My father was in the military and my father did that for me when I was fifteen, and I went to one high school which, after moving around all my life - it really does stand out in my mind that it was important. And then when I got the same opportunity for my son, it was kind of like, 'Yeah, this is the right thing to do'. And my mother-in-law's health wasn't very good, and anyway - we moved back here in '88."

We had been gliding across the soft sand of Smith's Point, but are now bouncing roughly along two ruts pretending to be a road through a

Sunrise over Esther's Island

duned sea of beachgrass. We have reached Esther's Island. At 6:15 a.m. we park on a bluff commanding a view of Madaket Harbor. Below us, six or seven boats restlessly crisscross the slightly choppy, shadowy, blue and orange sea. A handful of boats are approaching us, and still more are heading to different areas in the harbor. A traffic jam of thirty or so boats are preparing to drag off of Jackson's point. Dave will blow an airhorn at 6:30 sharp, signalling the scallopers that they can drop their dredges. "Just to slow 'em down a little," are his words. Years ago, there were no time limitations. Dave explains that they are safety related.

Although we are sitting on Esther's Island, we have not crossed any water to get here. Ten years ago, we would have. Fifty years ago, a truck would have done just as well as today. In 1961, this bluff was known as Smith's Point. But that year Hurricane Esther blew through the narrow spit of land, severing the original point, which became Esther's Island for the next forty or so years. It was so close to the newly created Smith's Point that even a mediocre swimmer might have been tempted to chance a crossing, except for the roiling, intimidating tidal current just a few feet off the beach. The rips, currents, and shoals around Nantucket have sunk over two-thousand vessels in the last four centuries, and are to be treated with respect, if not outright fear. A swimmer may notice when a tide is running alongshore: it deposits them up or down the beach. But a tide running off a point always sweeps a swimmer out to sea. This is an important piece of knowledge in Dave's world, because in addition to his harbormaster and shellfish warden duties, he also manages the town lifeguards, who save dozens of swimmers each season from just such predicaments.

Hurricane Bob visited in August, 1991, followed by the Halloween 'no-name storm' which spawned Sebastian Junger's book, *The Perfect Storm.* A vicious and frigid December nor'easter struck in 1992, and in 1996 Hurricane Edouard raced through. By the time they were finished with Nantucket, a dozen or so homes had been washed into the sea off of 'Sconset and Cisco, millions of dollars of damage had been done, thousands of cubic yards of sand had been moved, and the gap between Esther's Island and Smith's Point had been closed. The names are now interchangeable.

A map of Nantucket readily shows where it is headed. Its shape is slightly like a comet, for the same reason. Its south and east facing bluffs are being cut away by pounding winter storms at the rate of several feet a year, any protrusion ground down and smoothed out. The leftover sand is swept around the ends of the island, to create new shoals off the points, with some accreting to the gentle sloping beaches of the north side. It is essentially being swept to the northwest. But less sand is being deposited than is being swept away, and by one estimate, Nantucket as a physical place will cease to exist in less than six hundred years.

Dave is looking through binoculars, his preferred method of keeping track of the fleet. He can tell who is who by the shape of their A-frame or the color of their boat, and rattles their names off for my benefit.

"Billy Medeiros. Karsten Reinemo. Billy Medeiros looking at *me* with the binoculars," he chuckles. "Herky Stojak. Frank Day right there and his wife Lavon. Back behind him is Pud Dunham and his brother Danny. There you've got John Kittila. Danny Holdgate. Kenny Aguiar. There you've got Howie Whelden. Blair Perkins. They were real good yesterday. Normally on the first day they're a little bit more antsy, but they were pretty good yesterday. Yesterday they all raced over and sat out here. Second day, they know they're here, the scallops are here, there's a little less urgency. They made nine hundred bucks last night, so life is good."

One boat breaks from the pack and approaches, presumably to satisfy curiosity as to whether it is indeed Dave's truck watching them. "Frankie Day, always in a hurry," Dave says. "Frank does everything just like that. He fishes hard, he plays hard." I ask Dave about the scallopers' unwillingness to divulge who they sell to and for what price, and he confirms that they have been very enterprising and creative in developing diverse markets, thus keeping the price up. "But you know, I've got to say for the most part they really are loyal. And just from my perspective, just an outsider looking in - I know how tired I was yesterday afternoon, watching these guys, and keeping track of everybody and making sure that nobody got hurt, nobody needed to be towed in, nobody got too many scallops, whatever. So I know they're tired. So, come five o'clock, when it's time to sell the scallops, I would just want to go to the same guy I've been dealing with, dump the scallops off, get my check or get my slip, and call it a day. Rather than going around and saying - 'Okay, how much you gonna give me, how much you gonna give me, how much you gonna give me?"

Dave continues. "This is the other thing that's important about bay scalloping: these are not the only guys making money. The openers are making money, the boatyards are making money, the engine mechanics are making money, the culler is making money. Of course the fish buyers and those guys are making money. The guy who sells outboard parts and the guy who sells dredges and the guy who fixes dredges and the guy who does fiberglass work. There's a whole group of folks that have work because of this. A lot of people get a month's worth of work here, doing one thing, and then another thing for a couple of months. I have a couple of foster kids, and I tried to explain that to them last night. In Nantucket, that's the way people have survived forever and ever. Yeah, it's nice to be able to go to work at nine and come home at five and collect a big paycheck, but you know on Nantucket, that's not normally the way."

Dave had told me that the purview of the Marine Department has expanded over the years, now encompassing not only the shellfish and summer boating needs, but also the lifeguards, water quality, environmental issues, beach access, and habitat conservation, including the nationally debated piping plover rescue effort, and law enforcement. Under that category, regarding scallopers, the main issue has been enforcement of the growth ring regulation, and the age-old problem of scallopers taking 'extras', or more than their limit. Penalties for scallop-related infractions range from a 'day off' of fishing for the first infraction to loss of license and/or monetary penalties. He chuckles in telling me a story about two guys that had fished for years, and all the wardens knew that they were getting extras, but could not catch them. "We'd watch them from shore. Go to the shanties. I even boarded them in the middle of the night a couple of times. Never found a thing. It was driving me crazy because I knew they had them. You know how we got them? Someone called them in on a cell phone - another scalloper. Said to check under the floorboards. Sure enough, they had a false bottom under the culling board where the inboard used to be. Stashed them under there. Took us years but we finally got them." I ask if there are scallopers who regularly test the limits.

"Oh, yeah. There's certain guys that you watch every year. There's certain guys you know are not going to take extras. Some of these guys I haven't seen all year. It's kind of a neat thing, you know. I don't want to say there's any love here, but there is a mutual respect. We have a joke, that their job is to catch scallops, my job is to catch them doing something wrong. I'm not out here like a maniac, counting scallops and that sort of thing. Why do I keep an eye on them all day? Because come four-thirty I want to be able to say - 'Howie Whelden was fishing right off there all day.' So if I have to go find him and tow him in when he's broken down, I got a pretty good shot at knowing where he is. And yeah, I'm gonna check his scallops too, and I'm gonna make sure they're not taking extras and that sort of thing. But a lot of it is the safety aspect. I was talking to the buyers from Bourne yesterday. They've got seven boats, okay? Seven boats!? We got seven right here. But, I think they know that we're not going to go home until they're all in. When we passed the by-law that they have to be at the dock by four-thirty, that was done for their benefit. And I don't remember having any complaints. Come February, at four-thirty, it's damn dark out here. And if I've got to come find one of them - and unfortunately I've pulled one out that drowned, I've rescued a couple that almost drowned, I've towed dozens of them in at the end of the day. So I think there's that respect. Yeah, I'm trying to make sure that they don't take any extras. But by the same token I'm going to be there for them if it goes bad."

"I just counted eighteen boats here in an area that's maybe three hundred yards by one hundred yards. You have to be a good boat handler to be able to fish in something like this?"

"Definitely. You're exactly right. You're towing forty-fifty foot dredge warps, six or eight of them, you've got to know what the hell you're doing. Especially out here in Madaket, because this water's a little bit harder to fish than in town, because it's so skinny. It's so shoaled up out here, so flat, there's not much room for error. Once that tide starts to drop, they can't fish out in that middle ground - there's just no water."

"How do they navigate?"

"They don't do sophisticated plotting and positioning and navigation, you know, with the GPS or LORAN. They use reference points on the beach. They just say - 'Okay, you line up that house and that phone pole and that's where the scallops are.' But these are the best of conditions, you know. It's warm, there's no wind, there's not much tide running. This afternoon, or tomorrow, when it's blowing and raining, it's a whole different scenario. It's hard to keep the boat on the tow, keep everything lined up. And not only do you have to know about the fishing end of it, but you also have to know about how not to screw everybody else up. And don't think for a minute anybody's going to explain it to them. Nobody's gonna say - 'Hey, let me show you how to do that!' It's not gonna happen."

"It takes two or three years to gain some respect?"

"It does. And it's a tough racket to break into. You'll see guys that are just starting out. They stay off to one side. They go out late so they're not in anybody's way. Because somebody said something to them at the Stop and Shop like - 'If you go across my tow again I'm gonna effin' kill ya.' You know, something simple like that. But they fish just as hard. They may not catch as many scallops. It's tough. But you know, as far as the safety aspect of things, they help one another. My fishermen are just as good a fisherman and just as hard a fisherman as the guy who goes thousands of miles offshore. You know. They're good fishermen."

The magical hour of six-thirty arrives, and Dave gives a blast from an air horn. The scallopers throw their dredges overboard and go to work.

I remark at how fast rumors spread and get embellished, telling Dave that yesterday, within half an hour of Mike, Neil, and Tracy following the anonymous scalloper over to Grey Lady Marine to measure his catch, word was buzzing through the fleet that he was supposedly being 'chased' by the state fishery people.

'Probably didn't have a license,' speculated one.

'Maybe he didn't,' mulled another, then thought better of it. 'No, he had to.' It seemed beyond comprehension that anyone would try to get away without a license. Then followed the inevitable fun-making that such an incident invites:

'He is so half-ass. He comes down to go to scalloping half the time when we're just coming in. He's got his oil pants on - the whole ass is out of them. His jacket is all ripped, one doesn't have the arm on it or something. And it's raining! Y'know, he don't know.'

'It's too bad. He's a nice enough guy,' recalled an old fishing partner of his. 'He was good! You know, he was dependable, he would show up, he'd catch fish. He really liked it.'

'He's a daddy now, I think.'

'Huh?'

'He's a daddy now, isn't he?'

'Well, I don't know. I think he is.'

'Is he married?'

'Probably not. I don't know.'

'Where the hell's he live?'

'I don't know. I don't think he knows.'

"Oh yeah, it's just that fast," says Dave. He puts his unlit cigar in his mouth. I ask him if he has always smoked Friscos.

"There's a strange story about those cigars. I was on an island in the Aleutian Chain called Attu, which is as far out as you can go before you get to Russia. They had an enormous naval station there, they had submarines, they had all kinds of supplies and fuel and everything for all the ships during the Second World War, and they had a big naval air station there. Anyway, after the war the Coast Guard took it over. Put a LORAN station there in about 1950. Went from a contingent of many thousand down to twenty-five Coasties and three dogs. So they had this lake up in the foothills, and a pipe that came out of that lake to the station and provided us with fresh water. Come to find out this pipe was asbestos pipe. We'd been drinking this stuff. There was a big push and

Pennel Ames and Hau Truong backing in.

barrett pier

It is barely eight-thirty a.m. and Barrett Pier resembles New York's famed Fulton Fish Market. Three boats are offloading and several others are snaking single-file down the channel toward the dock. Three trucks are on the pier, one has backed down the concrete ramp below the pier, and people are moving every which way in every imaginable vehicle, tossing lines, carrying bushel boxes, hitching trailers, backing up trucks, docking boats, rowing dinghies. I half expect an ultralight aircraft to fly over, or some daredevil to parachute in. No one stays more than a minute or two; it is a breach of etiquette to linger at the dock.

college fund

Lavon and Frank Day arrive at the dock. Frank is the scalloper Dave Fronzuto says is always in a hurry. While Frank gets the trailer and backs it down the ramp, Lavon tells me that their daughter, Rachel, is going to college next fall.

"It's going to be a special season," says Lavon. "This will put her through college. And then tomorrow we're going to take Hunter, our eight year old son. He opens now - just started this year. But we took him out one day last year, and now he really wants to get out there and do it. This year we'll take him on all his days off, because he loved it, he really loved it."

"So he's going to be another generation scalloper?"

"Oh, absolutely! I think Frank started opening when he was about Hunter's age. He and his cousins opened scallops in his grandfather's shanty - Philip Marks. He opened for Donny Holdgate senior as well, and he started going out from there. It's the same old story. He goes out with someone, makes enough to get his own boat. I started going when I married Frank, so I've been doing it for about seventeen years. The year we were married was the year we bought the boat."

"Do you both go in the morning, or does he pick you up after the kids are in school?"

"Oh, I get up with him! We're lucky because Rachel is old enough to get her sister Morgan, who is nine now, and Hunter off to school in the morning. But when they were all little, my family helped out. I would bring them over to my mother's or my sister's house the night before, and they would spend the night with them. They know the scallop season's here, they're just so used to it. They're born knowing this. It's just natural to them."

Lavon and Frank are natives, as is her mother. Her father had been a navy man stationed here. There are many similar marriages which resulted from the navy base activity during and shortly after the war.

"How long into the season

do you fish?"

"Years ago we used to go right until February, March. Now that I work for the town, I'm just taking my vacation time. They're really lenient with me: they let me take half a vacation day, and then go in and work the rest of the day. So hopefully we'll get a good couple of months in. Normally we go up to the beginning of January - and then Frank goes back to his job."

Frank comes down F Street with his truck and trailer, and just like Fronzuto says, backs it in, winches up his boat, and as soon as Lavon gets in the passenger side, takes off, quickly and efficiently. He assures me he'll talk another day.

Allen Tornovish is a face I recognize. He gives you the inspection sticker at Don Allen Ford once a year.

Allen's grand-nephew Jonathan Holdgate, also known as 'Tugboat'. Tugboat went out scalloping for the first time today with his parents, Jon and Lisa, but when he got cold, Allen brought him ashore.

Dana Howard

mystery

Dana Howard is a picture of tranquility in this chaos. He is sitting in a plank boat that won't start, with half an inch of water sloshing around the bottom, smoking a cigarette. He is waiting for the wood to soak up enough water to stop leaking, and trying to fathom how an engine that started perfectly well when he tested it yesterday won't turn over today. He simply views it as one of the mysteries of scalloping.

"I do it because I enjoy it," says Dana, a floor finisher by trade. "I do it to get out of that everyday rut on the island, you know, going to work. Here, you get to deal with weather, you get to deal with tides, trying to get on the tow, if anybody's on your tow. Getting up in the morning, the sun's coming up. You're not really locking into anything. So it's just nice. Just a feeling of being free."

In two days I have seen four instances of engine or donkey engine failure, and it seems apparent that this is a major concern. I help Dana haul his boat out of everyone's way.

women on the water

Linda Roberts and Max wait on the pier for Chris, Linda's husband, to pick them up. Max is a black Standard Poodle given to them by Dr. Voorhees, deliverer of hundreds of island babies, shortly before he died. Max, as it turns out, does not particularly care for water, but he does like being with Chris, so he goes on the boat.

"Scalloping was always in my family," says Linda. "My mom, Joyce, and my dad went out. My dad is Ken Holdgate Jr. and my great-uncle, Allen, was the warden for a long time." Linda's grandfather, Ken Holdgate Sr., was the airport manager, and fished when he could. The Holdgates are a large family, and most of them have scalloped. The name Donny Holdgate comes up often, as a number of people have fished with him at one time or another.

"We had a next door neighbor, Norma Hanson," Linda says. "One morning my mother would get us off to school, and the next morning Norma would, so they wouldn't get too tired. This was back when you could get six bags each! And they were full! That determined whether it would be a good Christmas or not. I remember my father saying 'That's a quarter a scallop' when us kids would fool around and crack them."

Of the 134 commercial licenses sold this year, sixteen went to women. Most cull, but a few have their own boats. Several husband and wife teams scallop. Some couples go out together first thing in the morning. In other cases, the wife gets the children to school and joins the husband later. A special rule accommodates this. A scalloper can start out on the water alone, but can get no more than three boxes before being joined by a partner. This prevents coming in with a single limit, then going back out and getting another double limit. In theory, the wife could go out scalloping first, while the husband got the kids to school. But it seldom seems to work this way in practice.

Chris scalloped with Herky Stojak for thirteen years, then got his own boat, from which he has fished for the last ten. "Last year was a pain," Linda says. "We went out the first day: three scallops in three-and-a-half hours. The tide was against us, the grass was against us, and there was nothing. We couldn't find anything."

"This year?"

"This has been great! Everybody's getting them. It's really going to be a boom for the economy. I think people that were maybe thinking about doing renovations, or whatever, they'll be able to now. Chris' old boat was really beat, so he took a gamble this year and bought a new one. It really worked out well."

Laura Lamb and Bruce Bartlett

Charlie Sayle - 'The Ernest Hemingway of Nantucket'. This year, scallopers are coming in by eight a.m. with their full limits. Even early in the season last year and the year before, it would take most of the day.

purity

Charlie Sayle runs Sayle's Seafood, one of the three year-round fish markets. He comes in with a man I do not recognize, but who turns out to be Michael Foley, the major scallop buyer who had generated so much excitement yesterday by offering thirteen dollars a pound. His family business, M.F. Foley, Inc., is one of the largest distributors of seafood worldwide. Mr. Foley is a tall and highly distinguished looking gentleman. If I were casting a Broadway theatre impresario, I would cast him. Mr. Foley comes to Nantucket to fish the first few days of every season.

"It's my vacation time," he says. "I come down here and fish with the old man - Ernest Hemingway of Nantucket." He tosses a gentle laugh in Charlie's white-bearded direction. "It's business, but it's pleasure," he says. "This is good for the soul. And it's nice to be in in two hours rather than ten hours, right Charlie?"

"Making up for last year."

Mr. Foley adopts a concerned and sincere tone when discussing the state of the fishery.

"Last year was a pretty tough season," he says. "Ten years, declining each year. Pretty good article in *Nantucket Magazine* on the potential causes. But it doesn't take rocket science to know that the correlation between house building and the number of boats affects the purity and chemistry of the water, in ways that we don't yet know. But the superiority of the Nantucket take is owing to, my father would say, the 'extraordinary purity of the water'. And we don't know what an equilibrium state is. As the article said, could be a storm during the spawning period that upset it, an innocent storm in July. But those hurricanes, the No-Name and Bob, definitely had a very harmful effect. They killed the eel grass, killed a lot of it. The eel grass, you've got to have it."

Charlie has now loaded his catch into his brand new, high-tech, refrigerated truck. Several other scallopers, sliding their limits into plywood boxes on open-gated pickup trucks, make no secret of their envy.

"Look at this! Where's the cold beer?"

"You are unbelievable!"

"You're getting better with age, I don't care what your wife says!"

"You are doing it Charlie! Eight o'clock and your limit, and your wife isn't awake to yell at you, man. You got a couple of hours of freedom!"

"That's why these guys fish, you know," says Mr. Foley.

Mike Foley of M.F. Foley, Inc., the major buyer of bay scallops in Massachusetts, comes to fish with Charlie Sayle every opening day. He says that "The superiority of the Nantucket take is owing to, my father would say, 'The extraordinary purity of the water'."

John Kittila rowing in

brewing storm

Geff Folger backs his masonry truck down the narrow road toward the ramp at Barrett Pier, negotiating the trailer to avoid six cars parked along the side. He is preparing to haul his boat, the *Eastern Glow*, out of the water. It is a nineteen footer with an Evinrude 110 engine. Talk has begun to circulate about an approaching storm, packing seventy mile-an-hour winds, easily enough to swamp a boat left on its moorings, or drag boats off their moorings and toss them against each other or onto the shore. Although this is near hurricane force velocity, gales are common on Nantucket, and people are used to the routine of battening down the hatches and securing the lines.

Geff hand-hauled his scallops today, which means no donkey engine, just good old elbow grease to pull the ropes towing the heavy dredges in. He is talkative, willing to divulge his secrets.

"Well - you don't go fishing in the big mucky, muddy places that the guys with a lot of power go fishing in. The scallops are a little smaller - you're getting forty pounds and not forty-five or fifty, but at the same time you tend to get done a little quicker. And I'm not greedy, forty pounds is plenty for me. If I get thirty-eight, thirty-nine, forty, forty-one, I made what I needed to make today."

Geff is a mason, which is notoriously hard work, and a typical day later in the season might start at six a.m. and look like this: "I'll take my scallops, drop them at the shanty. Then go to Myles Reis and find out what time I can get concrete, then I'll either go home and eat a quick lunch, and then go pour concrete. Or, I'll go back over to the shanty and open scallops for a little while, and then go pour the concrete." But, in what is becoming clear, the more scallopers I talk to, this week is special. Geff explains. "I'm trying to get this week for just scalloping. Get it under my belt, get the feel for it. I don't want to go back to real work. I don't like real work," he says cheerily.

"Is there a name for this kind of boat?" I ask.

"Scallop boat." Geff tells me the engine is new because he blew out his 88 horsepower engine two or three years ago. "Some guys replace their motors every couple years with something new. I'm big on second hand motors. I don't feel so bad if I blow up a fifteen hundred dollar motor. I'd feel terrible if I blew up a forty-five hundred dollar motor. And actually, it's my third year on this one."

The boat has good lineage. Bruce Bartlett had originally built it for himself and then sold it to Les Wilson. Geff got it from Les in trade for building him a chimney.

Like Laura Lamb, Geff didn't fish much the past two seasons. "Last year

I went for a couple of weeks. I kept the boat on the trailer and every once in a while I'd go out and get a box or two, if it was too nasty to do anything else, or too windy or too cold."

"Do you haul this out every day?"

"I've got a mooring over at Warren's. But it sank a couple of years ago on me, and once that happens you get scared. There wasn't supposed to be a storm and then the storm came. By the time I got out to do anything about it, you couldn't get near the boat. All you could do was sit there and watch it. I sat there and watched Jeanetta Jones on the Weather Channel smile and tell me how the storm wasn't going to hit Nantucket, when it's blowing around seventy. She said it's going about three-hundred miles south of Nantucket, so you won't have to worry. As my boat sinks."

The cranking noise of a winch becomes the backdrop to this conversation, as Geff hauls his heavy boat up. He will decide as he drives it off the ramp whether to take it all the way home or park it on the road leading to the Barrett Pier, as many do.

The main talk around the pier continues to be about whether the gale forecast for tomorrow will arrive. Several other besides Geff have taken their boats out of the water. Some plan to see how it goes. I am told that some scallopers fish different places during different weather conditions, and haul their boats onto trailers in case they want to go into town tomorrow, depending on the direction and strength of the wind.

"Anything can happen," says one.

"Yeah, and it probably will," agrees another.

Geff Folger hauling his boat out of the water ahead of the approaching storm.

Simple and small, Rick Kotalac's Brant Point Marine building next to town pier is a gem of artistic detail and craftsmanship.

boots

Storm or no storm, I am going out on the water with Rick Kotalac tomorrow. Rick has been fishing for twenty three years. He also owns Brant Point Marine, the primary boating supply store, which occupies an ideal location right next to the Marine Department at the foot of town pier. The store was designed by Rick and built by Bruce Killen, a well-respected and well-liked builder whose father, Sid, was a well-known scalloper. Bruce had solidified his reputation by being the contractor on a 'This Old House' project that was big news on the island in 1996, right around the time Rick's store was going up. Although simple and small, the building is a masterpiece of artistic detail and exceptional craftsmanship. Like many old treasures, it is one that people will look at a hundred and fifty years from now and marvel at what craftsmen they were back in those days.

Marty Morse works behind the counter and performs various boat repair tasks. He is a touch over forty years old, with a reddish-brown handlebar mustache and a friendly manner. Marty's soft and direct delivery radiates sincerity and concern for the needs of the customer. Yet, he also seems equally concerned with honestly and meticulously enumerating the qualities of each item he describes, as though the item itself might have feelings.

When I ask him how he comes to be working here, he gives me to understand that whatever job he does, it will be something involving the sea. Eventually he gets around to telling me he was on the ship that located the *Titanic*. He doesn't actually say that, though. He tells me that he sailed for two years on the *R/V Atlantis II*. I get the impression that if I hadn't recognized the name, it would have merely passed in mention, and we would have moved on to discussing what gear I need for tomorrow. Since I do recognize it, and am genuinely impressed, he acknowledges that yes, he was there.

Rick Kotalac and Marty Morse

Marty spent thirty-five summers and a couple of winters on Nantucket's neighboring island, Martha's Vineyard. "I lived in New Jersey, but I grew up on Martha's Vineyard," he says, echoing what many who have spent childhood summers on one of the islands feel - it can seem more like home than their real one. I have also noticed that people tend to think that there is some interconnection between the islands. In truth, it is quite the opposite. For one thing, it is difficult to get from one to the other. Most residents have only one experience in inter-island travel - to watch 'The Game' between the Nantucket Whalers and the Martha's Vineyard Grapes football teams. 'The Game' is in three weeks, on the Vineyard. That afternoon, cobblestoned Main Street will look like a ghost town.

I ask Marty what I will need to go scalloping tomorrow, particularly in light of the brewing storm. He takes me to the section of scalloping gear for all needs, and steers me toward the heavy duty Grunden oilers - which are really made of polypropylene, but still referred to as oilers.

"This Grunden stuff, pretty heavy duty. Hard to tear. This jacket - maybe a medium ought to fit you. Then trousers. Trousers are like a Farmer John type-thing. Pair of boots. Best thing to wear underneath this stuff is a bunch of layers. Wool's good. Pile's the best. Keeps you a little bit drier."

"These boots have any kind of a nickname?"

"Boots."

I expect a mischievous grin, sure that Marty has taken advantage of an obvious opportunity to needle the landlubber. But he has not. He is sincere, reassuring. “Sea boots, but most people just call them boots.” But he cannot resist a small hint of the expected grin before continuing. “These jackets are traditional fishing jackets. A lot of sailors go with Gor-Tex or something like that, a breathable fabric. Which is nice equipment but it’s five to ten times as expensive.”

I opt for the significantly cheaper brand, Helly-Hansen, because it can also serve as regular rain gear in the real world. Even at that, I have just dropped a hundred dollars. The good stuff would have been double.

“What about other stuff, like dredges? Do you supply those?”

“Not a lot. Guys try to recycle, repair them. Dredge equipment is pretty

sturdy stuff. We sell more of the supplies to refurbish the dredge - dredge rings, connectors, net bags that go on the back. And we also sell the donkey engines." Marty goes on to talk about the ravages of rust and salt water on fishermen's equipment, and the amount of time and effort required to keep up with it. I ask him if they do that kind of work in the shop, and he tells me that they do, but that most of the fishermen prefer to do it themselves. I theorize that the cost would be prohibitive of hiring someone at Nantucket wages to do routine upkeep on a boat, with tradesmen getting anywhere from twenty-five to fifty dollars an hour for skilled work. "Oh yeah," he says. "And also, it's just a matter of pride. They take care of their boat."

Martha's Vineyard and Nantucket will account for the vast majority of the nationwide bay scallop catch this year. Marty tells me how, years ago, they quit dredging the Vineyard's Menemsha Inlet so that larger boats couldn't get in, and that they banned overnight anchoring, all in an effort to keep the pond clean.

"So they're even more feisty than we are in a lot of respects?"

"Yeah. But they had a couple of real bad scallop years anyway, and the people in Chilmark spent a lot of money couple years ago to seed their half of the pond. The pond is almost a square, and the line across it goes diagonally between the two towns - Gay Head, which is now Aquinnah, and Chilmark. So on this side of the line all the seed scallops were put in. Well, the scallops all swam over to the other side."

"I've heard this, but I don't believe it. Scallops actually swim?"

"Oh, they cruise! I was working at this environmental institute over at Monument Beach, snorkeling near Pocasset, on the Cape. This buddy of mine and I were out collecting seaweed for Marine Biological Laboratory. He whacks me on the shoulder and points like 'look over there.' And there's twenty or thirty scallops in a group, swimming along - dadoot-dadoot-dadoot-dadoot. Like this." He shows me with his flattened fingers opening and closing against his thumb. "We're going - 'What?! What the hell is that?! A school of scallops swimming?!' And I had never heard or read anywhere that they do that. But they do. They were on a mission! Couple things are pretty funny. They see pretty well, for shellfish. They've got a lot of eyes - Euell Gibbons called them 'the blue-eyed scallop'. We had a big tank, about twenty feet long, five feet wide, for the students that we had there. Mostly high school students, some first year college Marine Ecology college students. We'd fish things out to show to them. One of the things that was kind of fun to do was to take one of the starfish and move it over next to one of the scallops. There's this delayed reaction - five seconds, four, three, two, one - and the scallop would give like a little 'huh?' And then - 'Aahh! Starfish!' And then cruise to the other end of the tank. It's a mortal enemy, and I guess the starfish are giving off some scent or whatever. But those scallops were cruising. 'Oh, boy! I'm outa here!' We asked the guys at MBL - 'You ever heard anything about a school of scallops swimming?' And they're like - 'Uh - no-o-o.' 'Well, we saw one,' we said. And they said - 'Oh - kay.'

the fish market

I wanted to clear up this mysterious business about who buys scallops from whom and sells to whom, and for what price. For years I had played volleyball with John Sudarsky, and the volleyball players gather afterwards at a local watering hole called The Muse. Tonight was volleyball night, so I saw my opportunity.

John came to Nantucket for the windsurfing in the late 1980's and stayed, working at a variety of occupations over the years, until he finally scratched together enough money to buy a former butcher shop which he turned into a place called The Fish Market. From 1991 to 1998, he was one of four or five people buying scallops wholesale. No longer being in the business, John speaks with surprising candor about those years.

"They were actually the beginning of the downslide of the fishery. The first year I was in the business was actually the last decent year that we had in the bay scallop industry. Guys were still getting their double limits well into the end of December that year."

I ask John about this mysterious and powerful Foley.

"Foley is a very old and respected fish family from Boston, and Steve Connolly at one time worked for Foley. Connolly is now the largest exporter of seafood products to the world. What these guys did is they basically trained the fishermen here to accept cash daily for their catch, and they convinced them that they were the only people who had a market for their catch. They basically set the price for scallops. If you were selling your catch to either Sayle's Seafood, or Glidden's Seafood, or Souza's Seafood, you were selling to Foley or Connolly. The first year I bought, I was selling to Connolly, and he would pay me thirty-five cents over the dock price. As soon as the airplane landed in Boston, the money was wired into my account. That first year when I was buying 23,000 pounds of scallops, maybe a thousand - fifteen hundred pounds a day, that's a lot of money to be dishing out on a daily basis. It was very comforting to know that the check was in your bank account the next morning."

"In the old days we're talking on the order of two or three hundred thousand pounds of scallops, and if I could have done 50,000 pounds at thirty-five cents a pound, then that's going to pay my rent all winter no problem. But once the bottom sort of dropped out of the fishery, that really was not a viable option. Not only for me, but for guys like Glidden's that have been dealing with Foley for years. So everyone started to branch out. There was really little choice in that matter."

"You were among the first to do it?"

"Some of the fishermen did it before I did on a limited basis. But I was the first of the fish markets not to automatically sell my catch to either Foley or Connolly. I thought there was a market that would pay more, and I didn't really want to settle for thirty-five cents a pound. What happened was, the first really horrible scalloping year was my second year buying. I had filtered information from the fishermen all year long. And nine out of ten guys think it's going to be over by Thanksgiving. Normally, I'd let the other guys set the price, but I had a guy that wanted as many scallops as I could get him, and he agreed to foot the bill. I offered seven bucks a pound, because I knew there weren't that many scallops. That was a dollar over what the other markets were offering. By the second day, every fisherman on the island knew it, and none of the other fish markets responded. So I ended up cornering the crop for that week. And as it turns out, by the end of the second week it was almost over."

"This was neither Foley nor Connolly?"

"Neither Foley or Connolly. He got most of the scallops, and I moved most of them. That was actually my best year selling scallops."

"Now, this is the kind of thing I would assume would create a little resentment. You're the new kid..."

"I don't know this directly, but I know that one other buyer used to sell to a very big account in Boston. Well, he didn't have enough scallops to fill his orders, and he lost that account. So that week he called me up, and I had been quoted in the newspapers saying I was selling them to New York. I had been selling some to New York, the Fulton Fish Market. So he called me up and said 'You're not selling f'ing scallops to New York, I know that. I'll f'ing kill you if you say that again. That's what he told me."

"Why was it so important to him that you not sell to New York?"

"Well, he didn't think I really was selling to New York. But the way that I sort of cornered the crop, he was pissed off. He lost a big account. He lost money. Now, I was stealing from him, basically. That's how he interpreted it."

"I've never had my life actually threatened in a meaningful way. What'd that feel like?"

"I didn't mention New York after that! I mean, I didn't think that he was actually going to follow through on his threat. But I was very nervous. I installed a padlock on my walk-in cooler. When I saw someone strange in the neighborhood, I called the police. You know, I was a little worried."

"Were you shocked to see the price at thirteen dollars?"

"Not really. Because I don't think it's going to last much longer than a month. I think that there's a lot out there in one spot, and that everybody's going to scoop them up fast. The Nantucket Bay Scallop at thirteen is a bargain. Basically there's so much more demand than there is supply right now, it's very straightforward."

Other volleyball players jump into the discussion. One is Jay Detmer, who runs the Nantucket Bake Shop, a longtime family institution tucked away near the part of town once called 'New Guinea' because it was settled by seagoing blacks, predominantly Cape Verdeans, and who formed a tightly knit community well before slavery was abolished.

"Who's going to pay thirty bucks a pound for a bay scallop?" asks Jay. With several volleyball players now under the influence of a beer or two, the discussion becomes quite lively.

"They're not. But they'll pay twenty."

"That's still pretty out there in my book."

"I know, but they've been doing it for the past five years. Hey, you guys are witness to the fact that you attach the name Nantucket to something..."

"I hear you there."

"The main thing about our scallops," says John, "Is that the typical bay scallop is like a count of a hundred per pound. Ours is a count of thirty to forty per pound. They're big! They're so sweet! They are amazingly sweet. I brought a bucket to the Boston Seafood Show and people said - 'You put sugar in them!' They were so sweet. They couldn't believe how good. If there's anything that epitomizes Nantucket, it's the bay scallop. It's our image."

"It is the best," everyone assents.

"Is the Cape having the problems we're having?" I ask. "The Vineyard?"

"We're twenty years ahead of the Cape. The Cape had no scallops for twenty years. The Vineyard had none for twenty years, but they're starting to bring them back because of their intense farming and management program. Here though, we're behind the curve a little. But that's really ahead of the curve, because development has happened more slowly here. It's like you're driving down the highway at sixty miles an hour and there's a bridge out. If you hit the brakes at the right time, you can stop and give yourself time to repair the bridge. Once you go over, it's all over."

"We're in the repair process now?"

"No, we're not. We're at the point where we're very close to going over the bridge. Once we're over that bridge we're going to be in the same position the people on the Cape are. It'll be twenty years before you see another scallop."

"It seems to me there are two entirely different things here. The harbor and Madaket."

"Well, there are. They're extremely different. But the way that everything in life is connected, I think that the harbor and Madaket are connected as well. I think if you have a healthy harbor, you're going to have a healthy Madaket. If you have a dead harbor, Madaket's going to have a much harder time being healthy. Madaket has much better circulation, the water's cleaner in there."

"Less big boats," chimes in Jay.

"Less boats, period," says someone else.

"Less boats, plus it's only one small area that's open into the ocean. One thing you've got to realize is that, what I've gleaned from the fishermen over the years, in the old days when it was healthy in both areas - one year they'd fish the hell out of town, and the next year they'd fish the hell out of Madaket. And then they'd come back to town, so it was back and forth, back and forth. Now - there is no longer the opportunity - you go where the scallops are."

"There's too many people fishing now, that's all," says another player.

"Well ... there's less people fishing now than there was in the heyday. There's actually probably twenty-five percent less fishing now than there was in the heyday, or less."

John got elected to the Harbor and Shellfish Advisory Board, so-called SHAB, after he left the fish business, and in fact designed and maintains their web site.

"What got you involved in SHAB?"

"The reason I got involved is that I really wanted to help preserve our fishery. I saw it decline. And I figured this is my best way."

"There's a lot of resentment towards Tracy Curley because she wants to close portions of the harbor?" I speculate.

"I think there's a lot of resentment towards Tracy because she feels it's necessary to close the fishery in a crisis stage. Once the resource diminishes what you have is a fear reaction. People are afraid that they're going to lose it. And they think the only people out there to blame are the fishermen, because they're the ones that have the most interaction. But traditionally, the commercial fishermen are the guys that are willing to make the sacrifices necessary to keep the fishery alive, always have been. That's why striped bass came back. Commercial guys said - 'We won't fish it for five years. We won't even catch a one.' And I think the same is true with scallops. The commercial guys, ninety-eight percent of them are willing to make the sacrifice. Ninety-eight percent of them would never take a seed, they'd be ashamed to ever do it. There's always a bad seed here and there. But what I've learned is what happens to the guys that are the bad seeds. They'll come to their boat the next morning - gone. Because the other fishermen know that they're cheating."

"And the guys that get their boats sunk - do they tend to get back on the water?"

"They get the message."

"So what got you out of the business?"

"If there were still scallops around, I'd still be in business. The last three years I had nothing to pay my mortgage in the winter when there were no tourists. And if those scallops were still abundant, I would have been around."

Scallop boat on a breezy day. Note the classic 'A' frame, which provides the height necessary for a 'donkey engine' to hoist the catch up onto the 'culling board'. The ropes, called 'warps', are of lengths roughly forty to fifty feet. Each pulls a 'dredge', which is like a rectangular basket, along the bottom. A single trip along the bottom, lasting from roughly fifty to several hundred yards, is called a tow. Most boats fish eight dredges at a time.

November 3

Sunrise: 6:14 Sunset: 4:34
Temps: Low 49° High 63°
Conditions@ 6:53 am: 61°, windy. Rain, heavy at times
Forecast: Windy, temps falling through 50°'s. Wind S changing to SW, 25-35 mph gusting to 55
Marine Forecast: Storm Warning East of Newport. SE wind 30-40 kts, increasing to 40-50. Seas 10'-20'. Visibility 1-3 miles in rain and fog. Minor coastal flooding likely along south of Cape Cod at high tide

belding report

The ferocious storm failed to live up to the hype, and most of the scallopers have gone to work. But it is too rough for Rick Kotalac to take a novice on board. Even inshore fishermen appreciate the power and fickle nature of the sea, which is why most everyone took precautions yesterday to trailer their boats home or secure them on their moorings. Some, like Geff Folger, have experienced the consequences of not preparing for bad weather. All fishermen know that once a storm has begun, it is too late. An unknown quantity aboard during a storm is one more variable that can lead to trouble, and trouble in a small boat can quickly escalate into disaster. My first trip on the water will have to wait until tomorrow. I turn my attention to the Belding report.

The Belding report is the seminal scallop study, written in 1910 by Dr. David L. Belding to advise the Commonwealth of Massachusetts on legislative policy regarding the fishery. It is an indispensible primer on the anatomy and biology of scallops in general, and the northern bay scallop in particular, as well as being a history of the Massachusetts fishery. In many respects, it remains the definitive study.

Drawing on painstaking research, using what were then state-of-the-art scientific techniques, Belding described everything from the shell - ('a beautiful, fan-like appearance' which consists of fourteen to nineteen ridges, averaging sixteen) - to the origin of the name (he theorizes that it derives either from the French 'escallope' or the Dutch 'schelp') - to its mating practices (hermaphrodism). The most important observation he made was tucked away blithely on page 10 discussing the details of the scallop's style of releasing a stream of eggs into the water first, followed, generally a few hours to a day later, by a stream of sperm, a method known as 'broadcast-spawning.' Belding wrote:

> "There are a great many chances in nature against fertilization of the egg, and the subsequent survival of the scallop larva, after fertilization, is most precarious. If it were not for the abundant supply of sex products, the race of scallops would soon be exterminated."

In Belding's day, only forty species of scallop had been identified. The bay scallop of the Northeast was called *Pecten Irradians*, though some began referring to it as *Pecten Gibbus var. borealis Say*. Today, the accepted number of scallop species has swelled to four hundred - partly because new species have been found, and partly because known species have been subclassified. Classification remains an inexact science.

The report paints a fascinating story of the early life of a scallop - from its spawning period (when water reaches 61.5 degrees in nature, or 68 to 84 degrees in captivity), to the first cell divisions (the eight-celled stage occurs eighty-one minutes after fecundation). It reaches the blastula stage nine hours after fecundation, and becomes a ciliated gastrula an hour after that, capable of swimming by waving its hair-like cilia. Within two to four more hours it enters the trochosphere stage, and grows a flagellum, or feeler, to guide it along, a small mouth, and the beginning of a shell gland. Somewhere between the age of seventeen and forty hours, the scallop enters its free-swimming veliger stage, which lasts five to six days, and is the stage when embryonic scallops are commonly called 'spat', living near the surface.

Late veliger or prodissoconch (double shell) stage follows, in which the scallop begins to look like a shellfish, and seeks a place to 'set'. The dissoconch stage is marked by alternating periods of attachment, usually to eel grass or old shells, or crawling along the bottom with a foot-like appendage, typically when torn from the eel grass or voluntarily choosing to relocate. Attachment is by a thread called a byssus, or byssal thread. The final stage is called the plicated stage, and is marked by the formation of ridges on the shell, and the final development of the internal organs, including the extremely large reproductive organs. It now looks like a scallop, although it is only approximately one-sixth of an inch wide, by Belding's admittedly arbitrary definition.

Some key points mentioned by Belding are gleaned from fishermen's observations. The commercial range of scallops in 1910 seems to be from the waters of Massachusetts Bay south of Boston to the Chandeleur Islands in Louisiana, in waters from five to sixty feet deep. He notes that sets are most abundant in areas of heavy eel grass, one or two feet deep at low tide, but that cold winters often kill the young scallops in those areas. He also points out, importantly, that while a scalloping area may be relatively large, the local set varies from year to year, so the size of the catch often changes yearly. The natural barrier to range expansion is the rough waters of exposed coastlines. At this point in Belding's understanding of scallops, little differentiation is made between the northern bay scallop, southern bay scallop, or even the sea scallop, in terms of fishery management.

A historical overview of the Massachusetts and Nantucket scallop fisheries is also included in Belding's report. The natives probably ate scallops, but there seems to be no evidence that the European settlers had a taste for them. In fact, it is generally believed that scallops were regarded as poisonous in colonial times, and were often used as fertilizer. Scallop shells, however, had long been used as decorative or religious emblems.

Scallops had been fished along the eastern seaboard since 1858, caught primarily by one of three methods: hand-collecting; using scoop nets roughly eight inches across attached to various length poles; or pushers, which were rectangular nets approximately eighteen by thirty-six inches that were pushed along the bottom by an eight or nine foot pole. All of these methods were accomplished while wading. The first mention of a commercial fishery in Massachusetts was in New Bedford's Acushnet River estuary, at the head of Buzzard's Bay. Fanning out from there, scalloping was considered a commercially important industry in Massachusetts by 1872. Martha's Vineyard began scalloping in 1875 and Nantucket either in 1883, if you believe Belding, or 1881 if you believe *The Inquirer and Mirror,* Nantucket's newspaper of record, which noted that the first shipment of scallops off-island took place in 1881.

The 1874 introduction of the dredge, (pronounced 'drudge' on Cape Cod) revolutionized the industry. The first primitive dredge was a variation of the net at the end of a pusher, only attached to a rope and dropped overboard from a boat. Several at a time could be towed, initially behind sail-powered boats (often catboats), and later by motor powered boats. What scallopers now refer to as a 'tow' was then called a 'drift', with six to ten of the three foot wide dredges being towed. As time passed, variations on the shape of the dredges sprang up, often in response to the local bottoms. Chatham's was called a 'box' dredge or 'Chatham' dredge, and was used exclusively there, although a few cropped up on Nantucket. A second shape was called a 'scraper', which featured a triangular frame and was designed to dig into the bottom and cut through heavy eel grass, with room for extra weights if necessary. This was in use in Nantucket and Edgartown, the difference being Nantucket made its entire holding basket of twine, while Edgartown made the bottom of the basket of iron rings. A third variety of dredge was called a 'slider', in which the rail that drags along the bottom is angled up or made so that the blade hangs loose, so it will bounce or slide over obstacles, and is mainly used on bare or rocky bottoms. A fourth type, called a 'roller', was used only in Mattapoisett. The leading edge consisted only of a line of leads, and it was considered useful, even necessary, on a bottom that presented many rocks and obstacles that would otherwise snag or rip ordinary dredges. A steel-ringed version of the 'scraper' dredge is the one that has come into nearly universal use today.

Belding divided the fishery into four sections: the north side of Cape Cod, the south side of Cape Cod, Buzzard's Bay, and the islands. Most towns in these areas developed fisheries that supported a number of men through the winter. In 1899, however, all but two towns in Buzzard's Bay lost their scallops for unknown reasons. By 1907, however, those fisheries had recovered, and by the time of the Belding report, Barnstable, Orleans, Brewster, Wellfleet, and Provincetown were catching scallops on the north side of Cape Cod, while Chatham, Harwich, Dennis, Yarmouth, Mashpee, and Barnstable again, (it borders both sides of the Cape), were catching them on the south. The Buzzard's Bay fishery consisted of New Bedford, Fairhaven, Mattapoisett,

Marion, Wareham, Bourne, and Falmouth. But the Belding report saves its most glowing description for the islands of Martha's Vineyard and Nantucket:

> "In both islands the natural conditions insure the preservation of the fishery ... The rise and fall of the tides is slight, not averaging more than 2 feet. The variety of bottom mostly covered with eelgrass is found in all localities while the depth of water over the beds averages 10 to 15 feet rarely exceeding 25.
>
> In this section, no matter how scarce the supply may be elsewhere, the yield of scallops is fairly constant. Extreme scarcity and superabundance, so common in the other sections, occur here but rarely, and the scallop supply from this locality is considered the most dependable in Massachusetts."

Eel grass is the scallop's city. It is one of the reasons why Nantucket and Martha's Vineyard have long been Massachusetts' main scallop producing areas.

argopecten irradians irradians

I also drop into Tracy Curley's office in the new town building annex, diagonally across Washington street from town pier and the Harbormaster's office, to verify this phenomenon of scallops swimming. Tracy, in her capacity as town biologist, spends quite a bit of her time studying the harbor and the life in it, especially the scallop. Her desk is surrounded by filing cabinets, several pictures of her three year old son Dana, and various stainless steel and glass instruments of mysterious scientific utility. She tells me that most of them are used to test water for various types of pathogens.

Tracy always wanted to be a marine biologist. After Wheaton, she trained in Woods Hole for field work with the Sea Education Association (SEA), through Boston University. Her postings as a fisheries observer took her aboard a Russian trawler off the Atlantic seaboard, and then onto an American-Norwegian fishing boat in the Bering Sea. As as scalloper, she culled for Dan and Johnny Holdgate, and Eddie Backus, in the early 1990's.

Tracy fills me in on what has happened in the scallop world since Belding. The Nantucket Bay Scallop has since been reclassified as the subspecies *argopecten irradians irradians,* or northern bay scallop, which ranges from Cape Cod to New Jersey. Two bay scallop cousins, *argopecten irradians concentricus* and *argopecten irradians amplicostatus,* are found, respectively, from mid-New Jersey through both coasts of Florida, and along the western Gulf of Mexico down to Colombia. A fourth cousin, *argopecten irradians sablensis,* lived off of Sable Island in the North Atlantic, but has been extinct for several hundred years. *Argopecten gibbus,* the Florida Calico, while smaller than a bay scallop, is actually thought to be a sea scallop, found in deeper water, and probably the original stock from which the bay scallop descended. To determine subspecies, scientists study such factors as number and shape of ribs; color, size, and shape of shell; reproductive habits; and serological traits. Uncertainty arises because it is not clearly known which differences are genetic or which are caused by environmental adaptations.

I had entered Tracy's office knowing that scallops are bivalves, assuming this meant that they possessed two of some kind of valve, perhaps to filter water. Not so. It actually means that they have two separate shells, hinged to open and close. The word valve, as used this way, comes from ancestral English, denoting each section of a folding door. Also, the word 'scallop' was more likely derived from the ancient Germanic word 'skal', an all-encompassing term for any kind of protective sheathing, than the Dutch 'schelp' as Belding mentions. Other cultures, unable to pronounce 'skal', softened it, i.e. 'schelp', or 'shell', as in English, or added a vowel such as the French 'escallope'.

All scallops are of the family *pectinidae,* which comes from *pecten,* the Latin word for 'comb'. Karl Linné, the great Swedish botanist who adopted the Latin name Carolus Linnaeus, presumably as homage to his passion for classification, was the first to identify and name scallops. Many others have since taken on the challenge. French scientist Jean-Baptiste Lamarck is credited with classifying *argopecten irradians irradians.*

Scallops inhabit every ocean and bay on the planet, and are universal in their appeal. Pilgrims visiting the gravesite of the apostle and fisherman James in Compostela, Spain, bought St. James scallop shells *(pecten Jacobeus)* as souvenirs, to such an extent that the Pope had to ban their sale. The french scallop dish Coquille St. Jacques is named after St. James. Scallops have long been used as architectural and clothing design features, sometimes laced on a coat of arms to indicate Catholic lineage. The ancient myth of Aphrodite, immortalized in Botticelli's 15th - century painting *The Birth of Venus*, features the Goddess of Love herself rising from a scallop shell. There is a book called *The World's Most Beautiful Shells*, by Leonard Hill, filled with sumptuous Peter Carmichael photographs of shells, exotic and rare, spectacular and stunning, every kind of land and water shell one can imagine. In this book are representative scallops from Indonesia to Iceland, Chile to China, and everywhere in between. And on the cover of this book, the most beautiful shell of them all: a red Arthritic Lion's Paw scallop. The internet hosts various scallop shell trading sites, with specimens listed by species, classifier, date of classification, size, and rarity of color. On one site, the largest shell was an *amusium ballotti,* found off Australia in 1861, measuring 107 millimeters. It sold for ten dollars. An Argentinian albino *chlamys madrynensis* was the most expensive, at a hundred and forty-five dollars. Even calicos, which wash up on the Florida shore by the thousands, fetch two bucks. Such is the reverence for the beauty of the scallop shell.

Argopecten irradians irradians rates no mention among shell collectors. The northern bay scallop is a workmanlike scallop, undistinguished in size or beauty. Carl Sjolund couldn't even give them away to dump on his friend's driveway. But among the world's gourmets, it reigns supreme. And the Nantucket Bay Scallop is considered the finest.

Different scallops species have different lifespans and reproductive strategies. Bay scallops do not mate - they are hermaphroditic. They release from 12 to 18 million eggs into the water when temperature conditions are favorable - usually in early summer. After waiting an appropriate time, usually a day or two, it is thought to avoid self-fertilization, they release a similar quantity of sperm. The term for this reproductive procedure is 'spat'. Although I have heard the term 'spat' used as both a noun and a verb (scallops 'spat' in early summer, or millions of 'spat' can easily be killed off by a freak summer rainstorm), I have never heard the process referred to as 'spatting'. In any event, if enough scallops are concentrated in an area, fertilization begins and a new generation of scallops grows and begins the search around the harbor for a home. The best guess is that only one or two eggs per million will survive to become adult scallops. A bay scallop lives roughly two years, and conventional wisdom says that eighty percent will reproduce only once in their lifetime. Tracy believes, and other biologists are in the midst of research they hope will prove, that far more scallops will spat a second time.

Conch egg case. Conch are one of several scallop predators.

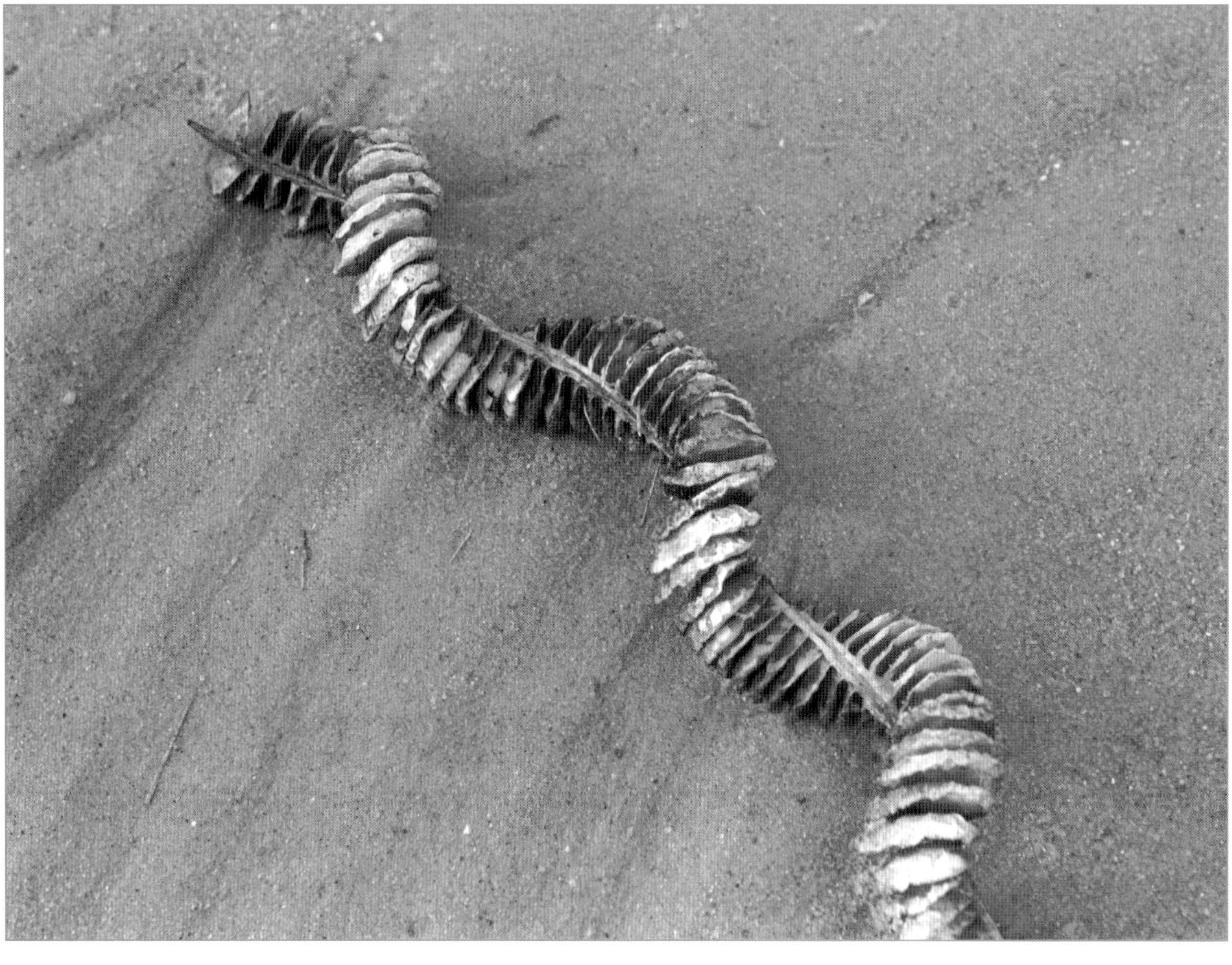

the survival quatrain

My interest in scallops began during my first October on the island. I found twenty or thirty scallops shells, all partially open, with the scallop gone. I derived great joy from arranging them in a semi-circle on a piece of driftwood, facing the largest one to look like a choir, and sending it to a friend to garner envy at my beachcombing existence. But I wondered for years what had eaten the scallops so neatly.

Belding described scallop threats in a chapter called 'Enemies', dividing them into 'Enemies of the Young' and 'Enemies of the Adult'. Enemies of larval scallops include other filter feeders, which literally suck the microscopic spat out of the water, but he noted:

> "The greatest destruction is from adverse natural conditions. Severe weather, storms, sudden changes in temperature, and water pollution may bring about destruction of the floating larvae."

Belding prefaces 'Enemies of the Adult' with the following:

> "The adult scallop has several natural enemies, both active and passive, and in addition is subject to the adversities of nature. *The effect of a hostile environment is especially detrimental since there is only one generation spawning each year, and its destruction would be disastrous.*" (italics mine)

Belding lists starfish as the number one active predator, though primarily a Buzzard's Bay problem. A starfish kills an adult scallop by encircling it with its tentacles and exerting a steady, strong pull until the scallop eventually tires and opens. The starfish then rolls its stomach outside of its body, and actually surrounds and digests the scallop.

Enemy number two is the inch-long oyster drill, which tries, as the name implies, to drill through the scallop and gradually suck the life out of it.

The third listing is called 'Passive Enemies', and includes a variety of growths such as oysters and worm tubes that can grow on scallop shells and either weight them down or bind their shells together. Other shellfish that compete for food are also passive enemies.

The fourth section is entitled 'Man', and reads:

> "While the main cause of the decline of the natural clam, quahog, and oyster beds is overfishing by man, the decline of the scallop fishery cannot be so considered. The scallop has a short life, hardly 20 per cent passing the two-year limit. So it does no harm to capture the marketable scallops which are over sixteen months old, as the scallop spawns when one year old and dies a

> natural death usually before it reaches a second spawning season. When only old scallops are taken, it is *impossible* for man to exterminate the scallops by overfishing. The capture of the seed scallop, *viz.*, the scallop less than one year old, which has not spawned, tends to the ruin of the fishery. As a rule, it is hardly profitable to catch the "seed" scallop, owning to its small size, but when the market price is high and scallops are scarce, it then becomes profitable. The present scallop law now defines a "seed" scallop and forbids its capture. By protecting the "seed" scallop, the State has done all that at present appears expedient to insure the future of the industry."

But since the Belding report was published nearly a century ago, new predators have emerged. Conch have always been present in New England waters and have always been scallop predators. All crabs are scallop predators, but green crabs, a non-native species that has spread throughout the Northeast, have recently emerged as a threat as well.

Scallops, like all mollusks, fend off enemies by building a shell which is difficult to pry open or drill through. But unlike other mollusks, they can also escape by swimming away. The ribbed design evolved in order to create a lighter shell to allow for swimming. They swim by opening and closing their shells and propelling themselves backwards. They can also turn by propelling jets of water to one side or another as they go. And with a ring of blue eyes around their shell, they can see danger and escape it. I asked Tracy where else a swimming scallop might be going.

"Oh, it's all pretty much food dependent, temperature dependent, light dependent. Light links in with the amount of food in the water. Different temperatures, different oxygen levels. They do move around - mostly with the current. They'll propel up and then they'll get carried with the currents. It's a chemical reaction. If they're feeling too stressed, then they will move out of the area. Clams can't move - they're stuck there. So if you have a low oxygen event, maybe they'll all die. If they're too crowded, disease can come in and wipe them all out. Scallops do have the luxury of quote-unquote 'swimming'. But they're not consciously saying - 'I'm going to go down to the Head of the Harbor today! And then back to Abram's Point!'"

I leave the office chuckling at Tracy's analogy. Scallops, with their blue eyes and their zig-zag swimming and their click-clacking, have taken on a personality in these last two days. But I have also come to realize this: they are merely out for survival, the same as any other species. And all species require four necessities to this end: sustenance, safety from predators, and resistance to disease are necessary for the survival of an individual scallop; and success in procreation is necessary for the survival of the species itself. The survival quatrain.

With that in mind, I look forward to my day on the water tomorrow.

"There are a great many chances in nature against fertilization..." - The Belding Report

Adult scallop predators include crabs, conch, starfish, and oyster drills when they are in the water. Scallops left uncovered at high tide will be picked off by gulls.

November 4

Sunrise: 6:15 am Sunset: 4:32
Temps: Low 42° High 61°
Conditions@ 7:53 am: 48°, light rain.
Forecast: Partly cloudy, wind WSW 24 mph
Marine Forecast: Wind W around 20-30 kts, increasing to 40 kts by late afternoon

on the water

I am driving west into a big black cloud. Cold rain gives the Madaket Road an eerie metallic sheen. I feel like I am in one of those digitally-enhanced car commercials currently seducing Madison Avenue. But, hopeful blue slivers light up both horizons north and south, and indeed it stops raining by the time I park and don my newly purchased oilers. Oilers are not worn against the rain anyway; scalloping is wet work, even on a dry day.

Rob and I are to meet Rick Kotalac at dawn. Rick, the owner of Brant Point Marine where I geared up yesterday, is taking us out on the water. This will be my first trip in a scallop boat, indeed one of my few trips in any boat, other than a small kayak or the large ferry which connects Nantucket to America. Nantucket being one of the great sailing and water destinations of the world, this is like living in Colorado and never skiing, or New Orleans and never eating.

Rick ties his boat to an employee's girlfriend's parents' dock. It is the kind of creative, word-of-mouth arrangement common to many endeavors here, be it housing, shop space, storage space, or boat space. As we depart the dock, Rick hands me a pair of long, thick rubber gloves, and suggests I take my oilers off before putting the gloves on. I do so, but subsequently realize that the rubber is so thick I cannot turn my tape recorder on. Knowing it to be a mistake, yet unwilling to believe it, I take my glove off. I will soon know the price of this particular folly. Several attempts to put it back on fail. The long, thick gloves cannot be made to slip inside the cuff of the oilers no matter how many methods are tried. Beaten, at last willing to admit that a scalloper who has been doing it all his life, and sells the equipment to boot, probably knows the proper way to put his gloves on, I take my arm out of the oiler. At that moment we accelerate into the channel, engine roaring and hull bouncing, and I find myself clinging to the A-frame like a drunk to a lightpost, for fear of being pitched overboard. Loathe to relinquish my life-in-the-balance grip, I cannot pull my glove back on, nor can I get my arm back in the oilers. My penance is to suffer my back being soaked by the spray we are kicking up, crystalline droplets suspended in mid-air against a bright rising sun.

At this moment, my predicament becomes completely irrelevant. Now past the docks and houses and bushes near shore, the horizon suddenly opens onto a sky full of mottled clouds, looming orange and black and purple and gray against the saturated blues and yellows and greens of ocean and beach-grass. I have hiked in every mountain range in the Continental United States, seen almost every waterway from the Everglades to the Boundary Waters to the Colorado River, driven the coastlines from Key West to Maine and from Mexico to Oregon, visited at least half of the National Parks, and even descended into a handful of caves. But this morning's trip out of Madaket Harbor is one of the most memorable I can recall. It is so beautiful it stuns.

The business of going to work quickly intrudes. At this speed, with the boat bumping and crashing in water chopped up by a stiff breeze, there is no hope of putting my oiler back on. I grab the dangling sleeve and hold it as best I can, firmly maintaining my death grip on the A-frame, and helplessly watching my brand new tape recorder get lashed with spray. I have no idea

Rick Kotalac

how long this ride will be, and am thankful that within four or five minutes we abruptly slow down. It is peaceful, despite the sharp westerly wind. After a few moments to take it in, I remove my jacket, put my gloves back on, and put my jacket back on over them. Noting the five-word, two-expletive phrase I have chosen to describe my feelings about this process, Rick comments, "Yeah, you don't want to have to take a leak out here."

Rick Kotalac has been scalloping for twenty-four years, which seems incongruous since he looks to be around thirty-seven years old. He is in fact forty-five. He often wears wire-rim glasses, but as I am quickly finding out, glasses on a scallop boat are constantly covered with salt spray. Today he is either wearing contact lenses or does not need glasses on the water. His boat is a twenty-two foot fiberglass Mako with a Honda 90 outboard motor, and we had been doing fifteen or twenty miles per hour. Rick, in addition to owning his supply store, is one of the better known scallopers. A genial and easygoing, direct spoken guy, he had offered to take me out on the boat with him. He also mentions a book by Peter Mathiessen that I am familiar with called *Men's Lives*. It is about the commercial striped bass fishermen who have worked the waters off Montauk Point on Long Island for three and a half centuries, and who have slowly seen their way of life come to an end through gentrification of their communities, dwindling of their fish stocks, and increased regulation brought on by sportfishing interests. The book was written in the mid-1980's. "But he talks about it like it's a thing of the past," says Rick. "These guys here are still doing it. It's going on today, right here."

I ask him if we have slowed down because this is where we are going to fish.

"I just hit bottom," he says. Rick tends to speak matter-of-factly, letting the words themselves, rather than the inflection or the tone, express his thoughts. Hoping not to betray my unease at being stranded in a small boat on a windy day, several hundred yards from shore, wearing heavy clothes I can never hope to escape from, being a lousy swimmer, and possessing an irrational fear of water, I ask if this is a common occurrence.

"Yeah, it's real shallow out here. If the tide had been four inches higher, we'd have made it no problem," Rick allows. I find this answer strangely reassuring. At worst, I figure I can stand in chest-deep water and wait to get rescued. We make slow progress on the shoal for the next five minutes, Rick poking at the bottom with a four-foot gaff. In a few moments we are on our way again. I gather that most everybody cuts across the shoals to save time, since most of the guys in the fleet know every contour in the harbor and have fished each shoal and mudhole many times. But occasionally the bet doesn't pay off. I am also reminded of Dave Fronzuto's comment of how the water in the middle of the harbor is so 'skinny' when the tide drops. I am beginning to understand that fishermen must have mental maps that allow them to see the bottom of the harbor through experience and memory, to render it a knowable, tangible entity, whereas I look at it and simply see water.

"People ever get in serious trouble ?"

"Well, you got your boat sinkings every year. You get a real rough day, you know, guys are pushing it out here, and a lot of sea water coming over the stern. All these boats should be designed for scalloping. They should have a higher stern. Might not happen so much with this fleet. This is a real hard-core fleet, the true fishermen this year. Because the last two years have been terrible. You have to cough up the two-hundred and fifty dollars in March, so a lot of guys who used to go for two weeks didn't get their license. They're kicking themselves now."

"So if you sink, does Dave come out in his Sea Ox and get you?"

"Well, it's usually the other guys."

"The fishermen have a good relationship with the wardens?"

"You know, if you're not a jerk, they're fine. You've got a pretty rough group of guys here."

Just off Esther's Island - which appears somewhat lower from the water looking up than from the bluff looking down - Rick turns his boat away from the wind. He carries the typical four dredges on each side of his boat, neatly folded on plywood flats at the stern, a yellow coil of rope tied securely to each.

"Most guys throw their dredges by hand," Rick says. "I did it that way for twenty-three years." He pulls back on a lever, and to Rob's and my amazement, the white flats angle up and all eight dredges slip overboard and disappear into the sea, the coils suddenly paid out and taut, trailing behind like yellow marionette strings. Rick calls this device his 'sissy-rig.' "I'm always coming up with these gadgets to make life easier out here, keep it more interesting," he says, a twinkle of pride creeping into his expression.

Going with the wind as we are, an illusion of standing still is created. But we are in fact moving, and pulling the dredges slowly behind us. Rick asks Rob if he has been commercial fishing before, and Rob recounts a tale of his single attempt, with a friend whose boat first broke down, then got trapped in the ice. 'Talk about Mr. Know-Nothing and Mr. Dipshit,' he caps the story. While we are laughing, Rick pulls on one of the tow ropes to gain some slack, and throws it over a pulley on his A-frame. He then winds the rope once or twice around a smooth, shaft-like protrusion on his donkey engine, which I can now closely examine and finally understand to be a small engine which turns a winch. The donkey engine emits a particularly abrasive, high-pitched whine, but hauls in the first dredge in a matter of fifteen or twenty seconds. The dredge is brought roughly to the level of the water, but Rick must then lift it manually the rest of the way until it swings, dripping water and seaweed, over the fiberglass culling board which cuts across the center of a scallop boat like a three-foot wide counter. Flipping the full dredge, which looks to hold about thirty pounds of scallops and seaweed, Rick empties its contents onto the culling board. This is a test dredge, to see if we should tow longer before hauling in. This one appears to be a good haul, so we begin the process of 'culling', or separating the seaweed, old shells, and 'seed' scallops from the keepers.

Rick grew up on the island and his family dates back many generations. "To the whalers," he says. He is referring back to the late 1700's through mid-1800's, Nantucket's Golden Age, when the lucrative and romantic whaling industry was centered in Nantucket. I seem to recall that Rick has a brother named Steve.

"Yeah. He likes to sportfish. I get a lot more of a thrill out of this than sportfishing." In this case, the disdain in Rick's voice is unmistakable. "I'd rather pick up a hammer than sportfish."

Brant Point Marine came into being after Rick's first year scalloping, when he realized he could buy supplies in New Bedford for half the price that they were available here, and that there was a potential market selling to other fishermen. Initially he was only open October through December, and operated at several different locations, but has become more-or-less year-round since settling in at his new building.

More dredges are hauled in by the donkey engine, lifted and swung by Rick over the rail of the boat, and through this process an enormous pile of scallops has materialized on the culling board. Rick gives me a wire basket and instructs me to work from the edge of the pile, pushing the 'shack' off the board and into the water as I go. Shack is basically anything but scallops - old shells, eel grass, crabs, or anything else dredged up from the bottom. If the dredges contain mostly eel grass, it is called grass rather than shack. My job here is to separate the good scallops from the shack and to throw back any seed scallops. In looking for the growth ring, I can see immediately where it might be difficult to separate adults from seed. All these scallops look basically alike to me, and I am trying to work quickly to whittle down this enormous pile. Rick assures me that these are

all adults.

Rick pulls up and neatly folds his dredges, and we begin to move again. He had only brought us over to Esther's Island to show us what it was generally like having to fish through all the grass and muck on the bottom, and we are now going to a prime area. We talk a little bit about the weather. I have often noted that when Nantucketers talk about the weather, it is not necessarily the kind of idle, conversation-starter people normally use it for. Weather has a genuine impact on how they will go about their day, make their living, and can even put them in physical danger. I remark that more Nantucketers per capita probably watch the weather channel than any other place in the country, as evidenced by the fact that many people, like Geff Folger, can refer to most of the anchors by name.

"And they're wrong forty percent of the time," says Rick. The observation seems earned by hard, experiential evidence.

We have now reached Jackson's Point, where most of the fleet has been fishing, and drop the dredges. Within minutes Rick is bringing them in, or 'hauling back' as it is called. He is pleased as the dredge swings up over the rail and hangs above the culling board.

"Now we're talking about all scallops, no seagrass. Get a picture of that, Rob! You'll maybe never see that again in history!" The second dredge comes up, again with nothing but scallops. "Ha ha!" rejoices Rick, still pleased with a good haul after twenty-three years. "I'm gonna empty it quick Rob, you can get 'em flying in the air!"

It truly is a remarkable sight, and one gets the feeling the entire bottom here is 'paved with scallops', one of two terms I have heard often in the last few days. The other is 'ass-deep'. And for the last four days the giddiness has been contagious. Largely because the fishermen have seen the catches decline year after year, and the work has gotten harder and harder for those who stuck to it.

"What made it a such good year this year?"

the inboard," Eddie grins, "But we didn't have any inboard. The compartment was below the culling board, and we had weather flaps on the culling board, so you could never see anybody going underneath it. We had rubber mats down on the boat - you see a lot of rubber mats because it's a hell of a lot more comfortable during a long day out there. So we just roll back the rubber mats, open that hatch, stash the bags down below. There's always a lot of salt water down there anyway. And nobody would be the wiser. But the wardens know people are out there taking extras all the time, and they love nothing more than to climb on board your boat and help you unload your scallops. They're looking around. They're checking for seed, they're checking boxes, too. I mean, some wardens can be lenient, some wardens can be real hardasses. Not everybody's fishing with new boxes."

I tell him about the anonymous scalloper who was told to get new boxes. "His boxes were all broken up. He had twice as many scallops in there as a regular box would hold."

"Of course!" Eddie laughs. "Our boxes would be like that, too. We always got accused. We used to tie our boxes together with string, they were so broken up. One of the tricks was to heat them up and take a stick, wedge them out so they're almost round. And some people would intentionally go and break them. Ours kept snapping because we had a bolt on our mast, and to pack the boxes, we'd slip one side on that bolt and stretch it with both hands. The next person would just shuffle them down. You do that a couple of years, your box is going to break somewhere. Those boxes are so much better for getting scallops than a brand new box," Eddie laughs.

"They don't seem to be giving too much of a hard time about crowning the top of them off," I say. The regulation says that the scallops should be level with the top of the boxes, but is largely ignored.

"Right, as long as the crowns aren't too big. There were a number of years when they were threatening to take a stick and run it across the top of the boxes. Some people literally threatened - 'You take that stick and run it across the top of my box and I'm going to hit you with it.' But we knew for years and years that the wardens wanted to catch us. They used to hide out up there in Pocomo Bluff with binoculars. They mostly watch you from the land, once in a while they'd go out in a boat. They had all these different spots, but we knew where the spots were. When Tony Marks retired, he always said his one regret was never being able to catch us. Which was actually pretty cool."

"We would always unload at night. We'd take our legal limit down to have them opened. Then we'd wait until it got dark, and drive down to Children's Beach where we were moored up. I'd stay on the beach and watch for anyone coming. It would be very calm, very quiet out there. Sometimes you'd even pad the oars. Still nights, you're real worried. Mal would row out to the boat, unload the scallops, and row back in, load them in the truck, make a beeline out of town and down to the shanty. People would drive by, I would hit the sand, just lie flat amongst the boats there. Middle of January, freezing. Malcolm would be totally quiet on the boat. Sometimes you'd have people stop down there and they'd be drinking or talking. You're like - 'Go away - get the hell out of here!' A couple times the steamboat came in. They shine that big spotlight, and we'd be rowing back in with a couple extra bags and we'd go 'Oh, no-o!'."

"One night Malcolm was rowing back from the boat, and I was there on the beach, and all of a sudden from the wrong direction these car lights just suddenly lit up on both of us. Mal was already close to shore and we were like - 'Oh God, they got us'. Turned out to be another scalloper who had seen us from over at the Steamboat Wharf. He knew what we were doing, and he decided he was going to come down and give us a good scare. He was laughing! We were shitting bricks and he was laughing!"

"It was a good little adrenaline rush, because you know you were doing something wrong, and you know you could get caught. It was part scalloping - you're out there catching your scallops, you feel real good. But you feel a lot better when you're getting away with extras. But in the last year we finally did get caught. Fronzuto told us they had actually boarded our boat a couple of times at night, and tried to find where we were hiding them, and they never could. It took somebody on another boat, with a cell phone, calling us in. We heard that from the wardens the next day. We got met by Jalopy - Albert Coffin. Jalopy never said a word to us, we figured we were good as gold. It wasn't until later that night Fronzuto got ahold of Malcolm and told him: 'Look boys, we found your spot. We met your extras. You're taking a couple of days off.' Told Malcolm how it happened. Just one of those things. We got ratted out."

I later asked Dave Fronzuto about the incident, and he chuckled.

"Oh, yeah. I've got a lot of hours invested in watching Malcolm. With him, you knew the wheels were turning all the time. You know what the funny thing was? I gave Malcolm a week off, I think. You know what he did for two of those days? He brought his kids over to my farm so they could see the animals."

November 5

Sunrise: 6:16 Sunset: 4:31
Temps: Low 44° High 60°
Conditions@ 7:53 am: 54°, sunny. Wind W 12 mph
Forecast: 55° to 60°, breezy
Marine Forecast: Small Craft Advisory. Wind SW increasing to 20-25 kts by afternoon. Seas 4'-8'

scalloping lesson

At 5:35 a.m., a sliver moon still dangles in the western sky ahead of me, even as the orange glow of dawn washes over the eastern horizon. I am on my way to Madaket again, this time to meet Bruce Cowan at Barrett Pier.

Bruce is already in his boat when I arrive, and has me take my place in the bow. 'Working the bow', as I understand, is another term for 'culling', my task yesterday with Rick Kotalac. We are underway the instant my boots touch the deck, slipping through the narrow channel, past the new Jackson Point bulkhead, still under construction, and into semi-circular Madaket Harbor, which opens to the sea on the northwest side. Bruce is unmistakably proud of his boat.

Bruce Cowan

"It's a twenty-four foot skiff, fiberglass," he says. "Kind of an unusual design, in that it's very long and narrow. It's got a good looking shape as a profile, but looking at it fore and aft, it's really kind of a canoe. It was built in '79 in Duxbury, when there was an energy crisis. You couldn't get gas for your car and it was even harder to get gas for your boat. So the guy made it very economical. There's only a forty-eight horse engine on here, which is not a big engine for a twenty-four foot boat, but because it's so long and lean it pushes it okay. You can run all day on a tank of gas."

"Being so skinny, does it -"

"Does it rock a little bit? Probably, yeah. I surf, so it's fine with me - I feel like I'm on my board. There's no such thing as a perfect boat design, because when you go in one direction, you lose in another direction. A deep 'V' boat is a great riding boat in a chop. But when you stop, that boat is a very unstable platform. A flat bottom boat is a very stable platform, but it pounds when it's underway. A soft chine, a rounded bilge - very smooth riding boat, but again, not quite as stable, more roly-poly. But I'm comfortable in this, and I've had it probably fifteen years."

"Speaking of which - a tank of gas. How much?"

"Ten or twelve bucks. But a lot of guys have these big engines, you know, a hundred and fifty horse, and they're ten-thousand bucks for a new one. This thing's like twenty-seven hundred bucks. Some guys'll change their engine every couple of years whether they need to or not. Like some people get new trucks that way, just trade 'em in and get a new one. Other guys run it until it's dead, and that may be anywhere from four or five seasons to ten seasons."

Several gulls are pursuing a duck. It dives to get away from them, but they set on it each time it rises. I have watched gulls all my life, and never knew them to be predators of ducks, or anything else for that matter, except shellfish, garbage, or dead fish.

"Oh yeah, I see them in the boat basin," says Bruce. "One day the mother will come in there'll be about five little ducklings swimming behind her. Next day there'll be three. Next day, one. Then - forget it. They gobble them right up."

As we leave Jackson Point and civilization behind us, symbolically at least, I reflect that this predatory behavior is, after all, a natural condition. It is as though, in passing this point, we begin playing by the rules of the water. But Bruce reminds me that scallopers are working people, and that carries its own set of rules. In this case, there was a price drop last night - thirteen to eleven bucks. "Little too good to be true," he muses.

Bruce confirms that the first week of fishing was as good as it gets. He even got so many that he gave his full dredges to someone else.

"Is that kind of camaraderie a common thing?"

"It's not uncommon. You can dump them back in the water - but if there's somebody there who wants them, you just go up culling board to culling board and push them off onto their board and save them a little work. It's just kind of a gesture."

"How long have you been scalloping?"

"Since the winter of '81. I came out here with the University of Massachusetts. My first winter here I worked at Nantucket Seafood down on Straight Wharf. I used to weigh up scallops for the guys coming in. So I saw these guys bringing these scallops in and making in a day what I was making in a long week. You know, I was a young kid, and thought - 'This is all right.' I'd grown up on the water in Scituate. I knew boats, I knew the waterfront a little bit. Didn't necessarily know the fishing business because I was too young. So the next year I got a boat. And I taught myself how to scallop. Made a lot of mistakes."

"What kind of mistakes?"

"Oh, just towing in the wrong spots, you know. There would be a cluster of boats, and that would indicate where the scallops were, a hot spot. But because I didn't have the experience, I didn't necessarily want to get in the fray, so I kept towing kind of around the periphery. And those were all the crappy areas. So once I got some confidence, a sense I could handle my boat, what was going on."

"How about if you were sort of an unknown like that, could you have gotten into that fray of boats, or is there some sort of etiquette?"

"Well, there is a little bit of etiquette - but the Nantucket fishermen are really generally speaking, a very fair, good bunch of guys. There are waterfronts around the world where it's just really ugly, competitive. So competitive that it's nasty. This is more - it's definitely competition, there's a macho atmosphere - but it's not nasty. Usually. Every once in a while there's an incident where somebody's boat gets damaged - but very rare."

"Something like would have a personal grudge involved, not really fishing related?"

"Could be, could be. Yeah. But probably more to do with fishing."

"So who are we out on the water with here?"

"That's Jim Crecca, David Perry, Jan Jelleme over there. This is Pennel Ames, Ken Kelley in the purple boat, Carl Sjolund is over there, you know Carl from the shanty. That's Karsten Reinemo from the Downy Flake."

All the people I have seen on the water have the expensive Grunden oilers Marty recommended, heavyweight full-length pants with suspenders and matching hooded jacket, usually orange in color, sometimes green or yellow or brown or even camouflage. I am experiencing oiler envy.

"They out in the same neighborhood yesterday?"

"Yeah, pretty much, but this place has been beat to hell over the past few days. What you'll see happening now is people will begin to spread out a little. Now that everybody's a little more comfortable with their routine again, with their boat, with their gear, with the area."

"Do you get tangled up in your prop very often?"

"Oh, every once in a while, especially in deep water. The dredges go down at a steeper angle, and sometimes when you turn, your forward line will go under the boat. You'll see guys reaching for their lines when they go to make a turn, bring them up over the railing, hook them onto a cleat or over something. Or once in a while somebody's not paying attention and you hook a dredge - dredge to dredge on somebody else."

Again I get the feeling that we are virtually at a standstill.

"We're still bumping along," says Bruce. "I can tell by that buoy against the shoreline. Basically you want to go as slow as you can, but not so slow that you're being inefficient." Bruce is scanning the horizon beyond me. "I don't really want to turn this way," he says, mostly to himself. I ask him what one is looking for when one is maneuvering.

"You're mostly just looking to make sure that the wind doesn't affect you too much, the tide doesn't affect you too much, that you don't turn too slowly, too sharply." A pair of hinged two-by-fours that look a little like wings stick out about four feet on each side of the boat, through which the tow lines run, making them less likely to wrap around the propeller during turns. "You have to think of not necessarily where you are, but where your dredges are. Because that's what's going to get hung up. They are a serious amount of resistance and your boat can literally swing on your dredges like a pendulum. And you've got to always look for what the other guys are doing. There are rules of the road but they don't necessarily always apply out here because not everybody knows them. And sometimes the wind is too much of a factor to let that come into play."

"What kind of rules?"

"Well, vessels to starboard have the right of way. Stuff like that."

"Basically nobody wants to run over anybody else's dredges?"

"Right. Because then you both stop, and the wind takes you, and then your lines could become a big, tangled, jumbled up mess."

I count twenty-two boats in sight, probably seven or eight within two hundred yards. And indeed, the nearest one seems headed across our lines.

"He can stop," says Bruce. "I've just got to make sure I don't go across Pennel's lines here. It's pretty tight. As long as you have guys that know what they're doing, it's not that big a deal."

"How far into the season do you go?"

"I'll go as long as I can make as much money on the water as I can on land. I much prefer to be on the water working than on land. All right - let's haul back and see how we do. We want to get right back on it. Because every boat that tows, there's that much fewer scallops. And with everybody in the same spot - once the scallops are gone, they're gone. They're gone for good."

When Bruce hauls back, I see that he has smaller dredges than Rick Kotalac's.

"The front dredges are only about sixteen or eighteen inches," Bruce explains. "The back ones are twenty or twenty-one. That's quite small. The biggest that you're allowed to fish with are twenty-eight, and I think only eight rings deep. So you can't have this long, enormous sock of a dredge."

Fishermen are notorious for quirks. Often what a layperson might call a quirk is really a comforting habit developed through years of repetition to make life easier. When Rick Kotalac culls, he tosses the scallop up in the air and lets the wind blow the shack and eel grass away from it. One woman won't wear orange gloves culling. If she and her husband have found a good spot, she doesn't want her hands culling quickly to attract attention. Bruce's quirk is that he smacks the end of the scallop on the culling board to see if it is hollow or if mud slides out. If so, it is a dead scallop, or 'dud' as he calls them, or 'mudder' as a dead clam would be called in Maine. He allows that it may be inefficient, but on the other hand, a scallop is worth about a quarter, so he doesn't want too many

duds taking up space in his box.

The air in the harbor is so clear that I can smell a cigarette, although the nearest boat is at least a hundred yards away. For about half-an-hour now I have done nothing but pick up a scallop, smack it on the culling board, toss it in the wire basket. Pick up, smack, toss. Pick, smack, toss.

Like all scallopers and many of my fellow Nantucketers, I have two or three occupations, one being house-painter. Painting also has a repetitive and boring rhythm, a rhythm that sometimes mesmerizes me into what my original painting partner, Cindy Jones, coined a 'painting coma'. A painting coma manifests itself in periods of long, involved daydreams, marked by little awareness of the the outside world or passage of time. My belief, not shared by all, and particularly not by my grade school teachers, is that this is a natural and healthy response to a lack of stimulus.

I realize that I am already in a 'culling coma'. Pick, smack, toss. Pick, smack, toss. Pick, smack, toss. The rhythm is barely broken by the occasional seed that gets thrown left into the water, instead of forward into the basket. On occasion, I sweep an armful of shack off the culling board. Twice I nearly sweep my tape-recorder off with it. I am truly grateful for a break in the monotony when I have to move a new pile of scallops in front of me. And it is positively a thrill when I fill up my basket. At that point I get to rinse the scallops by dunking them in the water a couple of times. It becomes an adventure, since the boat is running in a chop and I have to lean over the rail of the bouncing boat, swing the heavy basket toward the bow, lower it into the water, and then let it drift back toward the stern. It seems highly possible to lose the basket overboard, and as Bruce reminds me - 'You don't want to have to catch them twice.' Then it is

back to pick, smack, toss; pick, smack, toss; pick, smack, toss.

The gloves are probably second nature to Bruce, but they are still bulky and awkward to me. "You've got to compromise between something that's thick and durable but still comfortable," he says. Bruce culls at a rate of approximately ten to my one. He tells me that there is a lot of manual dexterity and feel to it, as well as eye-hand coordination. We slog through a lot of shack. Many insect-sized crablike creatures crawl over our catch. Bruce calls these animals 'krill', and I seem to remember reading that they serve as the essential food for most marine life, especially whales.

Hand-sized green crabs scurry away from the pile. Crabs, starfish, conch, borers, and gulls are scallop predators. But these particular green crabs are not native. They were introduced into the Gulf of Maine in the 1930's, probably in ballast from foreign ships, and began turning up as a problem throughout New England shellfisheries in the 1950's. Bruce kills every crab he sees with a swift hammerlike motion of his closed fist. A faster crab with more spindly legs nearly escapes him. This was an army crab.

"What I learned from some of the old timers," he says, "Is that you do much better with more short tows than with fewer longer tows. The fishing is faster, cleaner, you have a better sense of where you are, where you're getting them, where you're not getting them."

Culling coma

"Probably better for the harbor too?"

"Exactly. Heavy, heavy dredges tear up the grass. Now that's not to say at certain times, at certain places, you don't make really long tows. Sometimes you do. A good hard bottom, clean, and the scallops are very spread out, you will make long tows. And your dredge still won't be full."

"If you put your hand on the rope, can you sort of feel it?"

"Oh yeah. Later on when guys are fishing, you'll see them holding on. When it's banging, you can feel the vibration right up through the rope, that means your dredge is fishing on a good hard bottom. It's 'tending the bottom' they say. Whereas if it's just very smooth and it's sliding along, that usually indicates a grassy bottom, and the bottom edge of the bar, the bottom of the dredge, has grass wrapped around it and it's just sliding along, and not picking up scallops. The grass is stopping the scallops from coming in - it's just going right over them. Sometimes you have to make tow after tow. The first tows dig up all the grass, and later tows get the scallops. I'd rather have a lot of grass. Even though it's heavier, it's less frustrating than pawing through a lot of shack."

"Shack is different from grass?"

"Yeah, there's probably a layer of grass, a layer of scallops, then under the scallops you get all sediment, all full of shack, all debris."

"When you're culling, do you still keep an eye on the boat?"

"When I first started, I used to have to stop and cull. Most guys start out culling for somebody else - I didn't. I got my own boat. I went out there and made a lot of mistakes, but I learned. It's all about paying attention, making sure you're in the right spot. You get a range. For instance, you see that tower? You'd use that tower and that telephone pole and you'd say - 'Okay - this is where I'm going to throw my dredges. When I get up to this next range, that's when I'm going to make my turn and come back.' Stuff like that."

"Is it generally one big turn like this?"

"Some guys will tow in a straight line, figuring that the forward dredges clear the grass away for the bigger dredges in the back to really get all the scallops. Other guys fish in what they believe is a hole - where the scallops have collected during wind and storms. They'll just turn and turn and turn in a tight little arc, trying to scoop them up out of that hole. And some guys really tow for such a long period of time that they don't have any idea where they caught them. But they just do the same thing that they did the last time. That's one of the reasons that I like the small dredges, because when the scallops get scarce, you can pull the dredges up once in a while and see what you've hit where, and pinpoint them a little better. Whereas if you've got a big twenty-eight inch dredge, you're not going to want to pull that up by hand."

"There's one guy that still hand hauls, I hear."

"Bill Spencer. Naw, he's got an A-frame this year. Used to be all there was for a long, long time. I think until the '60's. Makes for a good fisherman. You really don't want to take long, long, long, long, tows. Because your dredge can only fit to a certain point. Once it gets full, nothing will go in it. You're just wasting time and fuel."

Scallopers will often handhaul a test dredge, as Doug Smith is doing here. Until last year, a man named Bill Spencer handhauled all the time. Well into the 1960's, most scallopers handhauled.

"So I guess the big debate going on right now is whether to close portions of the harbor. What do you think about that?"

"Well - this area was closed last year, and it was a good thing. The reason it was a good thing is because it was loaded with so many seed there were no adults. So if you're not going to get any adults, you don't want people walking around out here with pushrakes during family season stepping all over seed scallops. You don't want guys out here with big heavy dredges, tearing the seed up, tearing the bottom up, bringing the seed up in cold weather. But that's only when the seed is ass-deep. If you've got a mixture of seed and adults, the adults should really be caught. They're only going to die."

"Ever go the the SHAB meetings? That's probably where a lot of that policy is going to get made."

"Oh yeah. Toby Leske. Very good man. Very dedicated to the harbor, nice guy. Really putting in a lot of time and effort. You should talk to him."

"You don't want to have to catch them twice."

"So if on your last tow of the day, you wind up with a lot of extra scallops, more than your limit, do people just moor their boat and leave them in the dredges?"

"Well, no, because the dredges come up out of the water. Guys, lots of guys, have been caught taking extras. You know, they go back to their boat later, they've got an extra basket or so."

We have come close to achieving our limit, and it is before 8:30, so we go down the shore a few hundred yards to explore a new location. Bruce says that on Friday, people like to explore a new spot, so they'll know where to start up Monday morning. We only throw four dredges at this location, haul them up and cull them off in a few minutes, throwing a few handfuls overboard since we have gotten our limit. As we head back to Barrett Pier, Bruce very carefully puts his dredges up in order, on the culling board. First he puts the dredge down, then he wraps his rope in a nice coil on top of it, and then he folds his dredge back over the wrapped coil. He leaves his boat on a mooring at nights. I notice how carefully he maintains his equipment.

"I guess the last thing you want is to get these ropes tangled up when you're trying to toss them overboard. How often do you encounter mechanical difficulties?"

"Well, you try and stay a step ahead by keeping your gear well-maintained. If you neglect it, you're going to have trouble. It's your livelihood, and you've got to have the machines to do it. You don't have an engine, you don't go out. You don't have a hauler, you're probably not going to get your limit. If you do, you're going to wish you had your hauler. So the individual personality plays a big role. You take shortcuts, they can come back to haunt you. No different than the rest of your life."

November 8

Sunrise: 6:20 Sunset: 4:28
Temps: Low 35°....High 52°
Conditions: Clear

survival instinct

Two subtribes of the Algonquin nation originally inhabited the island they called Natocket, which roughly means 'the place that is far away in the sea'. Man had been in the area since the last ice age, and probably became isolated on the island as the Wisconsin glacier retreated and the sea rose. Ancestors of Nausets from Cape Cod occupied the east; later arrivals from Martha's Vineyard the west. These tribes were generally referred to as the Kauds and the Tomkauds, and are thought to have lived in several villages, at least one of which changed seasonally. The native history is generally known only through lore and myth. Villages were ruled by *sachems* and advised by religious leaders called *pawaws*. One ancient *pawaw* myth foretold that a great tragedy would befall the Natocket tribes, and as a result the beloved bluefish would vanish from the surrounding sea. They would return again only after the last native died.

Approximately 1,500 natives lived on Nantucket at the time of white colonization in 1659, but there may have been two or three times that many at the time of first contact, thought to be a 1611 landing by a Captain Edward Harlow, who planned to kidnap a native and take him to England as a circus attraction. Early visits throughout coastal New England resulted in ship fevers, against which the natives had no immunity, ravaging the villages and killing many. The Massachusetts tribes were decimated in that first generation, and by 1665 they were outnumbered 60,000 to 30,000 by the settlers. This led English-educated Metacomet, who became known as King Philip, and who was the son of sachem Massasoit, original welcomer of the whites, to wage the futile war which bears his name. But even had the natives managed to drive out the newcomers, they would never drive out their yellow fever, their typhoid, or their smallpox.

In 1763, 358 natives still lived on Nantucket. Sickness brought by an Irish brig visited the last remaining village, near Miacomet Pond, killing 222. Those who escaped had mostly been living as domestics in white homes, or were away at sea. The bluefish disappeared for the next ninety years. In 1854, Abram Quary and Dorcas Honorable, the last two true Nantucket natives, died within two months of each other. That summer, the bluefish returned.

I had spent the weekend thinking about ducklings being picked off by sea gulls. Sustenance, safety from predators, resistance to disease, success in procreation: the survival quatrain, as ancient and simple as it is inexorable: the fewer individuals that reach breeding age, the smaller the next generation will be, until the species finally disappears. The ducklings, although perhaps simply an example of survival of the fittest, became for me a prism through which to view new, unexpected knowledge. What defines fittest? And is mere survival truly the highest calling of life? Eons of evolutionary refinement becomes meaningless in an environment its host has never faced. Who knows how many species were on an evolutionary path that would have exceeded present day man, but were snuffed out by a microbe, a parasite, a temperature change, or simply a superior predator with a particular fondness for its taste?

In the New World, the native population, so thoroughly adapted to the American continents as they knew them, did not die from an individual inability to adapt to a changing society - they died from a genetic inability to counter a suddenly lethal environment.

As one moves up the evolutionary ladder, the strategies for securing the four necessities must grow more sophisticated. Evolutionists and creationists alike agree that socialization and diversity is the best survival path for man. In fact, researchers have recently discovered a human 'sharing' instinct, which seems, at first glance, counter-intuitive to individual survival, but which may in fact be a genetically-coded, long-term survival strategy. All that being said, those of us who have survived thus far are still, in reality, the product of stock that is most resistant to disease. We long ago nullified any real threat of extinction at the hands of predators or environmental stresses. Modern medicine has dramatically lowered the odds of early death by disease or genetic defect. In fact, the greatest statistical threat to pre-reproductive man is himself, through accident, murder, or suicide. How we adapt to this new wrinkle remains to be answered.

pack fishing

At our spaghetti dinner, I had mentioned to Eddie Loring that there were 29 boats around buoy number 22 out in Madaket.

"We would always avoid pack fishing," he said. "Pack fishing is probably the worst fishing you can do, because there's no rhyme or reason to it. You got these idiots towing right through a spot you just dumped all your shack into. And they're going to be dumping all of their garbage onto areas where there are perfectly good scallops, maybe a bed of scallops that hasn't been touched, and next thing you know, they're going to get all this garbage just dumped on them. The fishermen should be using that buoy to dump all their shack right around that, and be fishing all the areas around. But that never happens. It's very sloppy. I mean people are cutting you off, it's just a crazy zoo."

"Tangled up in your dredges?"

"Oh, yeah. Things like that happen all the time. If people try to come in, you'd try to cut them off, or make it look like it wasn't really worth it. People would ask how the fishing is, want to join us."

"And you'd say?"

"Oh, it was awful!" Eddie laughs, bringing from the oven some hot bread with butter soaked into it. It has grown dark outside and the warmth of the wood-paneled room seems to make the house much older than it is, as though it is retreating into history as we speak. "Malcolm had a temper out there. He'd tell them pretty much right out: 'Go somewhere else!' We'd pretty much try to stay out of the pack, try to keep people from fishing with us. We had a few tricks up our sleeve we would pull to keep people away. We had this one particular person who was trailing us for about three or four days. Malcolm said words to him, and that wouldn't discourage him."

"Literally across the water?"

"Oh, yeah. That definitely happens. This time of year, no. Everybody in the first week is pretty much gentlemanly to one another. Everyone's pretty much glad to see one another out there. But as the season progresses, and it gets tougher and tougher, people start staking out their claims and territories. Particularly because, if you find a nice spot, very easily one day the entire fleet could be right on your spot and you've lost it - it's gone."

"So the understanding is, unless you've been invited in, you should stay away?"

"It depends on the fisherman. There are those who, wherever the fleet is, they'll jump in. But there's some solitary people out there. Basically leave them alone. Let them do their own thing. That's a respectful kind of thing to do. It doesn't happen all the time."

"So you and Malcolm were known as solitary fishermen. So some guy comes along and follows you out to your spot. Malcolm says to him ..."

"Go effing fish somewhere else!" Eddie laughs. "Don't be jumping on my tow!"

"And sometimes the guy would actually stay, after hearing that?"

"Sometimes they would. Sometimes they'd swear back across the water. 'I can fish anywhere I god damn well want! It's a big free harbor and blah-blah this!' It's one thing if people come up and see you fishing, they throw a test dredge, see what the bottom's like, see what you're getting. They're doing the work for it. But there are people who see you fishing in a spot and they don't even throw a test dredge, they just come up and throw all their dredges behind you. As a good fisherman, you're kind of irate about that. We actually put a couple of new timers out of business. We had this one fella - about the third or fourth day that he'd been on our tow. He'd just come right behind us and dump all his dredges, follow us around. Malcolm had had enough of that. He went into the 'rockpile' - flipped all our dredges upside down so they weren't going to catch anything. So this guy threw all his dredges, expecting we were going to hit a hot spot. We're towing through it real easily, and his boat is going slower and slower, and the lines are getting real taut, and his engine's having a harder and harder time. We're laughing the entire time. We get near the end of the rockpile there - we just cut the engine, quickly pull up our dredges, and he's watching us. We circle back around so we're still in the vicinity, and he stops his boat, and he cranks up his donkey engine. He starts pulling in his dredges - and his boat is on an angle like this," Eddie slopes his hand at about a forty-five degree angle. "His donkey engine is just screaming. And his dredges are coming up all torn. They've got holes in them, these big huge boulders coming down on his culling board. To us it was the funniest sight, but to him he was really angry and discouraged. But he never followed us again. Folks tend to get the message. I think he fished a few more days and quit. But people threaten each other out there like that all the time. People threatening to run over someone's lines, cut their lines. We've actually had people try to run over our lines. That's happened to us on several occasions. People threaten to board other peoples' boats."

"Any names?"

"Probably prefer to leave names out of it," he laughs. "Lot of time it happens that when you're scalloping, you carry the anger over onto shore. That happens quite a bit. There's a lot of people who get angry on the water, but once you get on the shore it's over. You try to leave what happens on the water apart from what happens on the land."

"I've heard of people sinking each others' boats overnight."

"I've heard that being threatened. I've never actually witnessed that, although people have lost boats and people have speculated on things. Had one fellow out there one year who was known for jumping on peoples' tows. Was out in the midst of about a dozen boats and took on too much water. Finally ended up where his whole stern and half his boat was underwater, and he was sitting on the bow. And everybody kept coming by and saying 'Hey John, we'll pick you up, but we got to make another tow.' And the poor guy sat out there for about an hour-and-a-half with a dozen boats keep going around saying - 'Yeah we're going to come get you'. There wasn't any danger of him drowning and sinking, but nobody was going to let him on their boat. That was his last year fishing. He got the message loud and clear."

"But actually, what we like to do is have two or three or four people fish right around us. Basically you're fishing that one tow, and you all would agree on one spot where to dump - you'd run down the channel and fish and just double back to the spot. You'd keep that area clear. We wouldn't always fish together, but that's part of the camaraderie. You know that they go and they find their spots and they'll turn you on to it, and vice versa. You find your spot, there's plenty for a couple of boats."

"Who were some of the guys you'd like to see near you?"

"Doug Smith was one of them. Pennel Ames, Andy Roberts, Ricky. Those guys are all still fishing. The good fisherman and the old timers are the ones who get a lot of the respect."

"Whose names come to mind?"

"Herky Stojak. Eddie Stojak, but everybody called him Herky. Larry Miller, he's always out there. Oscar Bunting. We always admired Carl Sjolund and Jimmy Perelman. They were almost always the first ones in. Those guys, they almost always leave you alone and they're the guys you'd leave alone. And the people you'd have the most respect for."

"I'll tell you how good a scalloper Mal was. If there were scallops out there, he could find them. What makes the difference is being able to find your spots, and be able to stay in your spots. He knew just exactly where they were and how to stay right on them, tow after tow. This guy could take ranges - basically pick one spot, a house or a beacon or a radio tower, then pick another spot, and be able to keep yourself in line and parallel with it, watch how far you go. Get onto a spot, and narrow it down to a tee. Where you scallop and how you scallop depends on where you are. If you're in the shallows or the deeps, or in the channels. You've got to not only take your ranges, but you've also got to know the bottom. You've got to be able to feel the bottom and to know what your lines are doing. You can tell if you're dredges are full, whether they're fishing or not, or just bumping along. It takes a couple of good years of knowledge out there, working those areas. One time Mal lost his glasses overboard. A very expensive pair of glasses, prescription glasses. So he had to take the boat all the way back around, and he stopped. We looked down - it was shallow enough to see - and we were right back where we dropped them. This was March, and he had to take all his clothes off, and it was freezing cold, and you can imagine how cold the water was, like forty degrees. He dove in after them, and it was a split second later, like lightning, and he was not only up out of that water but he was on that boat."

"You could dredge up all kinds of stuff of stuff, too. In our early years there were old bottles out there, all kinds of different stuff. I told you about the bags of scallops people were planning on coming back for. Every once in a while we'd pick up somebody's dredge that they lost. Window weights are the best. You never throw window weights out. You go to the dump, all fishermen go to the dump, they go to the old windows, where they're tearing down old houses. They're always looking for old window weights."

"They're getting hard to come by!" I agree. "Used to be everybody had a box of them in their cellar. I remember looking for one a few years ago, I thought - 'I'll find one, that's easy. I'll go to the dump.' But, no. I was going 'Damn. Where do we get window weights?'"

"Fishermen have got them! It was always a great find when you could dredge up somebody else's window weights of the bottom. People'd use them for buoys, tie it to your buoy and toss it over. In fact that's probably the only place you're going to find window weights anymore is at the bottom of the harbor!"

Veteran's Day

Sunrise: 6:23 Sunset 4:25
Temps: Low 35° High 59°
Conditions: Windy, crisp

cemetery

Nantucket has many cemeteries. Far more people have died on this island than live on it, an oddly comforting hallmark of old, established communities. Growing up in suburbia, one does not realize they miss such things.

My first year-round home on Nantucket was near an ancient cemetery on New Lane, where people take pencil rubbings off the headstones. Some stand stately and white, others stark and gray. Still others are merely inlaid tablets with grass encroaching. Time and weather and lichens have softened the edges and smoothed the letters. Marble, so shiny and sharp and invincible when freshly placed, here betrays its vulnerability. Not unlike those resting below.

Another old cemetery lies next to the high school, across from the Stop and Shop. Still another rolling field at the top of Main Street covers several thousand Quakers. No one knows exactly how many, because religious stricture prevented them from leaving markers. Two small areas there are occupied by more liberal sects, who erected headstones toward the end of nearly two centuries of Quaker dominance. An Indian burial ground near Autopscot's ancestral village in Miacomet is rather vaguely delineated and remains mysteriously private, for reasons that I take to be respectful. And the largest cemetery is Prospect Hill, in the triangle between Milk Street Extension and Vesper Lane, bisected by Joy Street.

This cemetery, which is running out of room, has recently expanded to an adjunct location slightly away from town, on Somerset Road.

Today, on the gentle slope of Prospect Hill, a brilliant sun gleams through hundreds of tiny American flags, blown straight out, unfurled by a chilly wind. Enough, it seems, to make up a full regiment of Nantucket soldiers, soldiers whose memories are still fresh and alive, whose comrades-in-arms still live, and whose loved ones still recall.

Two full weeks of scallop season have passed, and most continue to get their limits by 9:00 a.m. About sixty boats still work Madaket; very few are fishing in Nantucket Harbor. The price of scallops has dropped to $10.50 per pound. *The Inquirer and Mirror,* the island's weekly newspaper, lovingly referred to as the Inky Mirror, has just come out as it does every Thursday. The front page photograph is of Kelly Mahoney, waving and smiling a dazzling smile, having just been named Nantucket's Junior Miss. And the state legislature has passed a bill forcing the Steamship Authority, an entity created to insure a year-round transportation lifeline to Martha's Vineyard and Nantucket, to operate a ferry from the Vineyard to the much more distant New Bedford. Such a ferry route once existed, but was shut down after years of losing money. The new move was justified politically as a means to decrease ferry traffic on Cape Cod, but is widely viewed on Nantucket as a boondoggle whose only purpose is to generate revenue for New Bedford's ailing economy, and which will probably benefit most the various politicos, lobbyists, and contractors who rammed the project through. Ironically, in the 19th century, New Bedford's new railroad line contributed heavily to Nantucket's decline in the whaling trade and subsequent economic downfall.

Also on the front page: in an unrelated move, the Steamship Authority raised its rates thirty-seven percent.

November 12

Sunrise: 6:25 Sunset: 4:24
Temps: Low 30° High 48°
Conditions: Light breeze

chain of custody

Richard Ray occupies an office next door to Tracy Curley's, at the town building annex across from the town pier. This is the Health Department, and Richard is the Health Inspector for the town. I noticed his name on various shanties' inspection stickers, and I assume that he authored the sanitation rules and regulations sheet that hangs in Carl Sjolund's shanty. I am curious to meet any government official who can write so clearly and directly. Richard is around fifty years old and looks very much like a distinguished middle aged government official, if you ignore the colorful cardigan he wears when he goes out, the longish hair, and the earring. A baseball bat which says 'The Enforcer' leans near his desk.

Richard inspects everything from septic systems to restaurants to scallop shanties. Shanty regulations began about fifteen years ago, because many people were opening in basements, living rooms, or in various other unregulated environments.

"Was that a town decision, or your own personal decision, or is that something that the state or some larger regulatory board covers?" I ask.

"Chatham had done it a couple years before. We felt this could only help to maintain the integrity of it, because it's such a unique species, as far as seafood is concerned. We don't charge a lot for the inspections, it's ten dollars. And we don't ask them to do that much, other than a clean shanty and some means of sanitizing the benches, and white fiberglass

RULES AND REGULATIONS TO BE FOLLOWED FOR SCALLOP SHANTIES

1. Floors free of cracks and uneven surfaces; graded to drain quickly.
2. Walls and ceilings smooth, washable, light colored, free from cracks and uneven surfaces and in good repair.
3. Shucking benches - tops and backs of smooth, impervious, nonabsorbent, nontoxic material, free from cracks and constructed so drainage is complete and directed away from all scallops. Corners shall be rounded. Material shall overlap the edges and seams soldered and sanded smooth. After each day's operation, all benches shall be either scalded with hot water of at least 180°F or washed with a sanitizer or chlorine solution.
4. Suitable number of waste containers shall be provided and no waste allowed to accumulate except in these containers, which shall be thoroughly cleaned at the end of each day's use.
5. Parts of utensils coming in contact with scallops shall be of corrosion-resistant, nonabsorbent, nontoxic, smooth material which will not readily crack or disintegrate. All seams and joints shall have a smooth surface for easy cleaning. At the end of each day, (knives used shall be so constructed as to seal the joint between blade and handle), all implements shall be scalded with at least 180°F water or disinfected with chlorine solution or sanitizer.
6. Water shall be of a source approved by Dept. of Public Health.
7. Toilet facilities shall be provided. They shall not be located in any room where shellfish are opened, shucked, stored, handled or packed, nor shall the door of such rooms open directly into rooms in which water closets are located. Doors shall be self-closing. If pit privies are provided, they shall not be located adjacent to the shellfish shucking-packing room and all toilet facilities shall be approved by the Director and/or the State Health Dept.
8. Lavatory facilities shall be adequate. Persons shall wash their hands thoroughly prior to commencing work and before resuming work after each interruption. A sink or wash basin with hot and cold water and soap with paper towels shall be provided conveniently located with respect to the toilet room. Signs directing employees to wash hands after each work interruption shall be conspicuously placed.
9. Adequate refrigeration, either ice or mechanical refrigerators, shall be provided to maintain temperature of 40° or less as soon as possible after shellfish are open.
10. All shucked meats shall be packed in new or previously unused containers of metal or such material as may be approved by the Department. These containers shall be clean and adequately sealed at time of packing. Prior to use, containers shall be stored in sanitary manner and protected from contamination. Metal containers shall be sealed by a friction-type cover. Sealing of other containers shall be adequate to protect scallops from contamination and approved by Department. The metal containers shall have impressed, embossed or otherwise permanently recorded on the side the permit number of the shipper, preceded by the state abbreviation and the words "Bay Scallops". Labelling of other approved types of containers shall be by a method approved by the Director. Bay scallop producers shall pack in containers bearing their own permit numbers.

It was no run-of-the-mill government official who composed this common sense, thorough, no-frills rules and regulation sheet, which hangs on the wall of Carl Sjolund's shucking shanty.

everywhere. A clean environment to open in. And it insures the product viability."

When I ask Richard if I can have a copy of the rules and regulations I saw on Carl's shanty wall, he tells me Carl must have typed them up and posted them himself. I mention how impressed I was with their brevity and thoroughness.

"That's Carl. That's 100% Carl, absolutely."

This year there are thirty-eight shanties. They are generally owner-occupied, one or two person operations. Many are owned by the fishermen themselves, like Carl, and some are owned by people who don't fish, but strictly open. The three seafood stores sometimes also rent shucking space on a small basis, to scallopers who will then sell to them. Richard inspects just before each season starts.

"We'd like to get there as early as three weeks before the season starts, but most of the guys don't have their shanties ready realistically until about a week before. Once in a while we'll get a straggler who on the last day will finally get the fiberglass to dry. It's amazing how many coats I can ruin in the course of a week by leaning against wet fiberglass."

Richard explains where a scallop goes after it is opened.

"As far as the product viability is concerned, we mandate locally that when people ship these scallops, they ship them in refrigerated containers - thermal blanket with ice packs. And we work with the airlines - they will actually refuse scallops that aren't sent appropriately - because they can be held responsible as part of the chain of custody."

'Chain of custody' is a term that I have never heard.

"The wholesaler buys them from the opener, who buys them from the fisherman," Richard explains. "And then there's the wholesaler on the other side, who buys them from our local wholesaler. And then it ultimately involves the restaurants, because when a restaurant buys shellfish of that nature, all the shellfish has to be appropriately tagged as to where it's from, and they need to keep those tags so that if there's ever a food-borne illness outbreak, if all things work as they're supposed to, we can trace that scallop right straight back to the individual who actually harvested it."

Oysters were a big part of the culture and economy of New York, going back to the Dutch days. Many of the early fortunes of Staten Island were made by 'oyster bedders'. The oyster flats all around the Hudson estuaries were fished out by the early 1800's, which gave rise to the practice of 'bedding' - that is, importing small oysters from the Chesapeake and letting them grow to maturity in New York's waters. In 1916, a typhus outbreak was traced, through a similar 'chain of custody' system, to contaminated sewage finding its way to the oyster beds. Oyster bedding ended, and clamming in waters around New York City has since been restricted to a few token, carefully monitored beds.

"Is that common to just shellfish, or is that all fish, or is that all fish and meat?"

"It's particularly common to shellfish - fish not necessarily. You have to know the individual who caught it. Shellfish, not only is it the individual, but the location."

"So when they sell them to the wholesalers here, are they telling them where they got them?"

"Well, the wholesalers know, and it goes on the tag. Right now about all that they're doing is Madaket."

"So all they have to say is Madaket? They don't have to say, for instance, Esther's Island?"

"Right."

"Ever had a problem in Nantucket?"

"We've had our problems, but it hasn't been with Nantucket Bay

Scallops. We had a couple of situations where we had some paralytic shellfish poisoning. It was individuals who ate scallops on the island. The crew of a boat that pulled in here because of storm conditions. We were able to chase it back to the absolute location."

"And how sick do people get?"

"Extremely sick. That's the only case we've ever seen, but it does happen. And it takes very few meats of whatever the shellfish is. It's a localized problem, it involves a particular seaweed or algal form in the area where the shellfish are. That's why the chain of custody is there."

"Does the Food and Drug Administration have anything to do with this?"

"Ultimately, the FDA will get involved in a case like that. But as far as the local scallops are concerned, no. Because you only eat the adductor muscle of the scallop. There would be a lot more regulatory control if you ate the other portions of that particular shellfish."

"Why would that be?"

"Whenever pollution is a factor, that pollution will affect the entrails of that littleneck, scallop, clam, because that's what the animal ingests. However that particular bacteria or virus does not get passed into muscle tissue. So even in contaminated beds, you can still eat the adductor muscle of the scallops."

"Nantucket has never had a problem, or has it?"

"Oh, we have some closed beds. Absolutely. Sure. If you were to draw a line from Eel Point to Warren's Landing, that whole section from that line towards the shore would be an unfishable area."

"For everything but scallops?"

"That's right. Portions of Polpis Harbor are unfishable at this point."

"Is this a recent development?"

"I think it's probably something that has been building up over the last five or ten years, and I suspect is a sediment situation, to a point where now it's really detectable. If it is a sediment situation and not a water flushing problem, it's going to be a long time before before we can remedy it. If it's simply a water quality issue because of a shore contaminant, that's short-lived and we can address that. But if it is in the bottom in the sands and sediment, it's going to be a big problem in trying to get it out of there."

"Do you work with Tracy Curley?"

"Tracy, the Department of Environmental Protection, the Division of Marine Fisheries, SHAB. Everybody's involved in a situation like that."

"Any consensus on the nature of the problem?"

"The finger gets pointed initially at septic systems. But I think there's more to the problem. I think there are time/temperature concerns. Where bacteria can flourish in a particular environment, that exacerbates the situation. I think waterfowl is certainly part of the equation here. Again that nasty word fertilizers can certainly be a part of the equation, in the sense that it causes activity in the water column that may be detrimental to the life of the scallop, or other shellfish."

"And these are being actively addressed by the groups you mentioned?"

"Absolutely. One of the things that has come out of the harbor water quality study has been a Harbor Protection District. We're going to try to make some changes that will affect the water quality of the harbor, which could bring back the eel grass beds, which hopefully will indeed bring the scallop back into the harbor. Now again, it's a small part, but land based remediation will be a part of it. Certainly we're going to have to do some dredging and other corrective measures, but I think it's within reach.

"This Harbor Protection District gives regulatory and enforcement powers to...?"

"The Board of Health can go within that district and say 'These are the activities within this land use area that are detrimental to the harbor. Here is what you shall do to remediate those situations, or make changes to, so that we can improve whatever water is running into the harbor and affecting the harbor water quality."

"Who pushed to get that designation happening?"

"It was myself, Tracy Curley, and other people who are now gone, who were a big part of this. The initial push was from the Nantucket Health Department, and the Land Council, Linda Holland. And we funded the Woods Hole Oceanographic Study, which was released about a year ago."

"So when did the ball get rolling on this?"

"About a year ago. When the final results came out, we moved very quickly to try to establish the zone. Had some problems at Town Meeting with the Attorney General not accepting our definition. He misunderstood the situation, we explained it to him, and the second year the zone passed without any problem at all. Now we have to go within that zone and do some work to try to make some changes. And those are going to be forthcoming. But it's a slow process, because we want to get it right the first time."

"In terms of other harbors, are we pretty leading edge on issues like this?"

"Oh, I think we're pretty pro-active, yeah. We're certainly not going fast enough for everybody. I want to be quick enough, but appropriately designed. I don't want to make band-aid approaches, quick fixes. It's a long-standing problem. It's not going to go away in a hurry, and I think it warrants a bit of a cushion of time to get it right the first time.

"What's happened to other harbors around the state?"

"You've got vast areas of closure because of the pollution problem. Bass River is almost a shellfishing disaster area. I don't know if the Vineyard's had any problems or not. There are other small estuaries on the Cape and in the Wareham area that have been closed. New Bedford has areas that are closed to shellfishing. In the case of Bass River, there are just a multitude of houses dotting the coast, within fifty or sixty feet of the water, certainly many more than we see on our harbor, they are probably contributing to a septic situation."

"I see a baseball bat that says 'Enforcer' on it, right next to your chair here. How often have you had to use that?"

"Only the threat, so far."

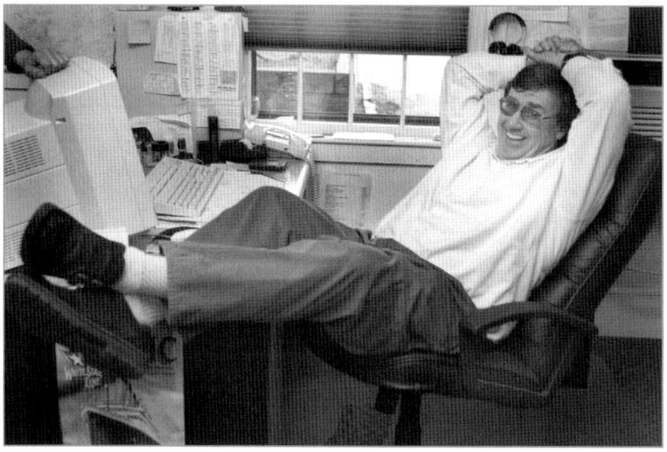

evie's shanty

The Downy Flake is a mid-island breakfast place famous for its take-out donuts. Located in the commercial center that has gradually spread out from the Stop and Shop, the Downy Flake opens its doors at 5:30 a.m. in the off-season, largely to accommodate the scallopers, and is immediately bustling. The woman who brought my pancakes this morning at the Downy Flake is Evie Sylvia. If you asked Evie what I had for breakfast in 1985 at their old downtown location where, prior to the unfortunate propane stove incident, the giant chocolate donut hung over the corner of North Water and

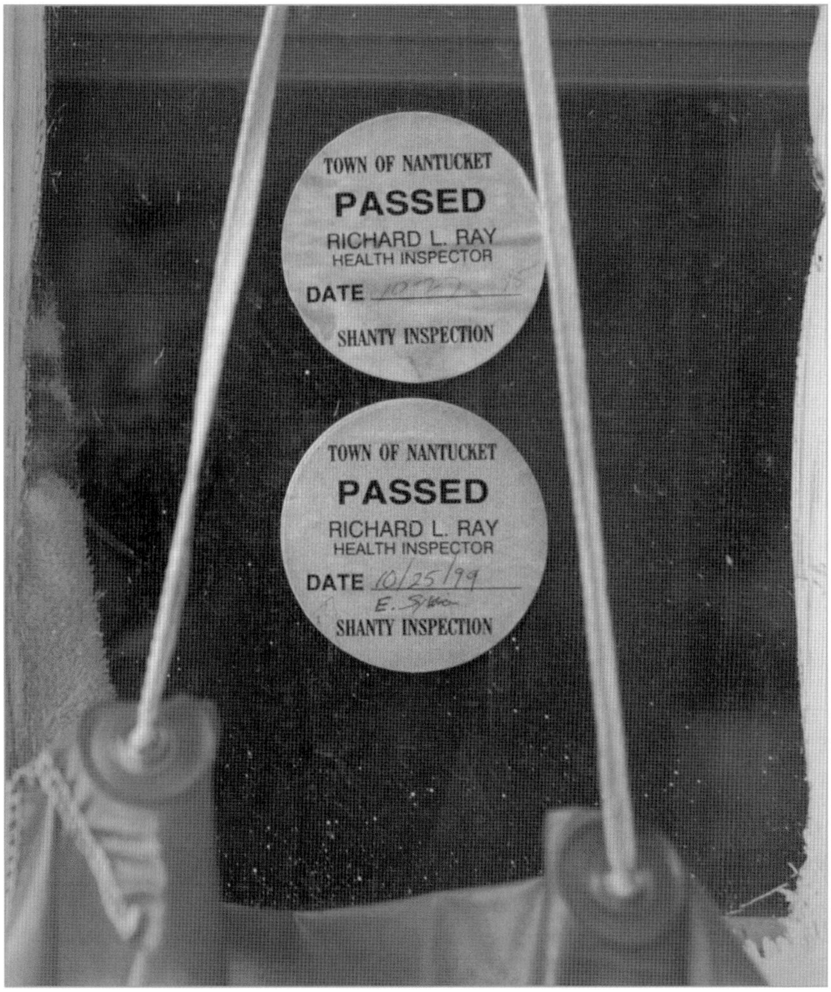

Evie's inspection sticker

Cambridge streets, she would answer without hesitation: 'French toast and bacon'. This year I learned that Evie also operates a shucking shanty, which is built into her garage.

When Dave Fronzuto and Paul Boucher checked on Evie's shanty, the garage door was down, and it was dark and empty, so he thought they were through for the day. Yet, he knew someone was shucking somewhere. The distinctive 'click-clack, click-clack', muffled in the dark, was a giveaway. The sound reminds me of an old *Star Trek: The Next Generation* episode where several members of the crew have vague memories of an alien abduction, and the aliens speak in the same click-clack voice.

"Paul and I were literally walking back to the truck, and I could hear the scallops," Dave said. "I could hear the clicking of the knife, so I turned back around and said, 'Y'know, they're flippin' here somewhere.' I'm like stealth fighter. I'm looking, and I've got my ear to the wall. Anybody who was watching me would have thought I was nuts. But then I see this other side door and, lo and behold! There they are, in there whittling."

When I knock on the same old beaded-board door with the wrought-iron latch, I find, much as Dave had, that it opens into a side room filled with light and activity. A white fiberglass counter stretches across the length of the shanty, with room for four workers. Evie stands next to the door on the left. Being right-handed, she takes this side because her mother, Irene Garnett, can then stand on her right, and they won't bump elbows. Mark Ray, a family friend whom Evie calls her 'adopted son', occupies the third spot, while at the far end is Evie's real son Elliot, who is the fastest opener. Kate Raphael, another 'adopted' daughter, has just finished shucking after a day at her regular job selling advertising for *Home and Garden Nantucket* magazine.

They are discussing 'beans', 'marshmallows', 'pillows', and 'Rice-a-Roni'. 'Beans' are tiny scallops, maybe two-and-a-half to three inches or so. Marshmallows are regular size Nantucket Bay Scallops, which are quite large by most bay scallop standards.

"Superbig ones are called pillows," says Irene.

"And this is rice-a-roni," says Elliot. Everyone laughs, as Elliot holds up about a one-inch scallop, a seed that didn't get culled out.

Evie and her mother Irene. Three generations open scallops in Evie's shanty.

"Elliot got his start opening scallops to pay for a broken window," Evie says. "We lived in front of the boatyard. Someone came into the shanty and said that there was a little boy on the fence sitting there crying. 'I know who that is,' I said. Well, a few of them decided to throw rocks. Elliot only threw one rock. That's what he says anyhow," she laughs. "So I went outside and got him and said 'Come along. You're going to be with me. You're going to pay for those windows.' Six years old. I had to put him on a scallop crate. His Auntie Gail and his grandmother would give him scallops, and Elliot would open in a margarine container. But he did pay the hundred and fifty dollars back."

"Hundred and fifty bucks?! How long did that take?"

"Few months," says Evie.

"Felt like years," says Elliot.

"But now he's the fastest opener in the shanty."

I ask Irene, Evie's mother, how long she has been shucking, since I know Evie has been doing it for forty-two years.

"Well, I'll tell you," she says in a very forthright New England manner. "My husband and I lived in Fairhaven, and my uncle lived in New Bedford.

And if you don't have a Fairhaven man on the boat, you can't go fishing in Fairhaven. So my uncle said to my husband, 'Why don't you come with me?' So we went over to where they were opening, I'm watching my uncle open and I say - 'That's so easy.' My brother says - 'Wanta learn? I'll show ya.' So he picks up a knife in his right hand - 'course I'm left handed. He looked at me and says - 'Forget it. I can't show you how to open scallops.' My uncle says - 'Come on over here, I'll show you how to open.' And for some reason I liked opening. Then we moved to Nantucket. My husband is from here, and just couldn't get work up there, so we moved back. I was carrying Evie then."

"Your husband was Henry?"

"Henry. He passed away - was the custodian at the town building for a long time."

"You might have run across him," laughs Evie. "He was kind of an ornery man."

"Ornery?"

"He was! My husband also fished, Elliot senior," continues Evie. "When we were dating he was scalloping. I told him he only married me because I knew how to open scallops. I remember the first time he took me out on the boat, and it was rocking and I looked at him like 'Is the boat going to sink?' 'My boat won't sink,' he said. Didn't sink. Was fine after that. We fished that year, twenty-nine years ago, up at Wauwinet. And the scallops, the inside of the meat, were as big as half-dollars. They were huge. It was just incredible the size of the scallops, I've never seen them quite that large before. Although these are close." Indeed, a number of these are very large.

"We lived in my in-laws' back yard in a little two room cottage," Evie continues. "My father-in-law used to go scalloping with my husband. When you just had to be able to see to go scalloping - you didn't have to wait until six-thirty. Elliot was usually out at the crack of dawn. Well, one morning on my bedroom window: Wham! Wham! Wham! I'm like - 'Omigod! What's the matter?!' It was my father-in-law. 'Okay boys! Time to get up! Time to go scalloping! You don't sleep in bed when there's scallops out there to catch.' Elliot's like - 'For Chrissakes Dad, it's only four o'clock."

I ask if I can bring Rob over to take some pictures.

"Make sure I'm not here!" says Irene.

"Oh, come on," cajoles Evie.

"It's not how I want to be immortalized!"

November 16

Sunrise: 6:29 Sunrise: 4:20
Temps: Low 33° High 42°

shab

Go to *www.shab.org* on your computer and a small image of Nantucket superimposed on a scallop shell will appear, overlaid onto an old island map, with the following greeting:

Welcome to the Website devoted to preserving the last surviving wild bay scallop fishery in America.

SHAB stands for Nantucket Harbor and Shellfish Advisory Board. Although the letters in the acronym make no apparent sense, the same cannot be said of the board. Composed of fishermen, biologists, and parties interested in the harbor, the main business of the board, which was formally chartered in 1976 after a decade of informal existence, is preservation of the harbor and the fishing in it, primarily the shellfish, and as stated, primarily the scallop.

A trip to their website is educational. A trip to their meetings, on the first and third Tuesday of every month in the town annex across the street from the town pier, is positively fascinating. It generally starts with some good-natured joking. The minutes are then read by Sue Mueller, the Marine Department's office manager, followed typically by an opening statement from Dave Fronzuto. Dave is not a member of the board, but is the person they advise and who implements policy as suggested by the board and acted upon by the town.

Tobey Leske is the chairman of the SHAB board. He is about forty years old, a clean cut, not overly tall or imposing man, someone who looks more like an accountant than his real trade as a licensed boat captain and a home inspector. I asked him about rumors that some New York City restaurants have been selling Nantucket Bay Scallops before the commercial season opened.

"It does exist. It is illegal. If they get caught they're going to be in trouble. I talked to a chef from there and he said yeah, they were getting some bay scallops from Nantucket."

"In his restaurant?"

"No. In restaurants he knew of, but he didn't want to give any names, which is understandable. I don't know what the scale is. I can't imagine it's very big. It is certainly something that we're going to have to look at for next year, now that we have confirmed reports. You know, it's typical. There are always people skirting the laws, that come here and they recreational fish and they take out more than their limit, hop in their private plane, and fly down there and sell it. It's difficult to catch them. I don't think it's a significant amount of scallops, it's the principle."

"You're concerned partly because people are breaking the rules, taking them and selling them. But what if they're not really Nantucket Bay Scallops? Would that be something that you would care about?"

"We'd care about it a lot. Anybody who knows scallops knows that Nantucket Bays are worth getting. Selling Calicos from Iceland or China or the Chesapeake or whatever, saying they're Nantucket Bays, I'm sure people do it all the time. That probably has more to do with the Board of Health, and shipping the product off-island illegally. We don't have the enforcement aspect, we don't have the people to enforce it. Dave's already stretched thin. Certainly if he caught somebody, he would fine them, and the fines can be significant. But that's not really our purview. Our purview is the management of the harbors of Nantucket. At this point we have more important issues that we need to deal with than a few dozen bushels going out to New York two weeks prior to the season opening. And also, Martha's Vineyard opens two weeks before we do, and they may be getting scallops from Martha's Vineyard and calling them Nantucket Bays. Or they are Martha's Vineyard Bays. That's entirely possible."

I ask Tobey if he ever scalloped - the answer is surprising.

"Never did. I'm not a commercial fisherman, not even much of a fisherman. But I'm a sailor. And that's my first passion, and I've lived on Nantucket all my life. Been around the waterfront my whole life. My family owns Cap'n Tobey's, (a restaurant on Straight Wharf), so I spent my days on the waterfront since my first job when I was maybe eight years old. Peter Gomes was the Harbormaster at the time. I used to meet him down at the dock at six-thirty in the morning and go out and pick up the garbage from the boats and deliver ice. Then I worked over on Steamboat Wharf for Ed Tarvis taking care of boats. I didn't get paid, just got to use the boats if I wanted to. Then I worked for a guy over on Straight Wharf, same kind of operation. You know, I've been on the water every summer of my life, and I just have a deep and abiding love for being on the water. There's no place I'm happier."

"Since I've been chairman, I've tried to make the focus of the board a more holistic view of the harbor. For many years it was just a gam for fishermen.

Which was fine, that's what they needed back in the sixties and seventies. That's really all that was needed at that point - to discuss the fishery, and the regulations for the commercial fisherman. But with what's been going on in the nineties and late eighties, with the boom in development and the consequent impact on the harbor, the issues we're facing are no longer just the scallops. The scallops are an indicator species. They're dying because the harbor's being affected by development. So there's no sense going out and buying two million scallop seed and throwing them into a harbor they can't survive in, as we used to do. You've got to make the harbor a place where the scallop can survive, and they may come back by themselves. I don't think we'll ever see the 100,000 bushels a year again, but we may see around 20,000."

"Was there a year when you noticed something had radically changed?"

"Well, when I got on the board I started looking at some numbers, and I put some graphs together." Tobey pulls two graphs out of a folder. "Here - 1980. 117,000 bushels landed. Last year they landed 6,800. I also have one that shows the building permits. And they're going in the other direction. Everybody talks about cycles - here it is charted out," says Tobey. "This is a cycle," he says, pointing to a three year up-and-down spike. "This -" he says, pointing to the twenty-year overall drop, "- is a trend." And indeed, the number of building permits issued has risen steadily, in inverse correlation to the downward landings of scallops. But Tobey is hopeful in that Dave's early predictions point to a season that may top 10,000 bushels. I ask if SHAB had an impact on that.

"We closed Madaket Harbor last year because there was so much seed. The fishermen hooted and hollered, but we decided to do it anyway, and this year we got a good set of scallops and everybody's very happy. Now we can say we did it once, and it worked."

"You guys can make that decision right in here?"

"No - Dave can close areas as he sees fit, but we usually discuss it first. We need selectmen to change the catch limits, but we have a very good working relationship with them."

Tobey tells me that a computer modeling study of the harbor is being done by Applied Science Associates in Rhode Island. The town funded it on SHAB's recommendations, and gave assurances that they will fund recommendations that might come out of the model.

"You can ask it anything - cut here, dredge there, raise the jetty, do this, do that - and it will come back and tell you what the effect's going to be in six months or six years. You can plug in any timetable you want, plug in any parameters, any time of year, any kind of activity, and it gives you a little fifteen or thirty second video. We saw some demos that they've done in other countries and around the United States, and it's amazing. For instance, now it takes up to sixty days for a molecule to come in through the Jetties, go up to Wauwinet, turn around and come back out through the Jetties. That's a long time, and that's one of the problems. The problem is not just the pressures from development, it is that the harbor is silting up and not flushing itself like it used to. So this computer model is geared to solving that flushing problem, hoping to decrease the flushing time, which will stave off the effects of development temporarily. And in the meantime, we're going after development too, but that's a more difficult process."

"We've had problems with loss of eel grass," Tobey continues. "We did a study down by Consue Spring to find out what kind of contaminants are coming through there, and how close is it to capacity. It comes down into those creeks and into those marshes, and as it goes through it filters out all the chemicals and heavy metals. Nutrients, chemicals, pollution are not necessarily being filtered out the way they should be, because it is very close to capacity. We did studies on the soil samples in the harbor by the outfall pipes and found there were indeed some heavy metals and some asbestos going in there, and so we're going to expand that and find out how far they've gone. Constant water quality monitoring and keeping good records of salinity and turbidity and temperature, establishing a general data base for the harbor. Woods Hole did to some extent in the early nineties with their report. What we're doing is adding on to that data, and getting a well-established database of information about the harbor. That's the first step. Then we can look for solutions to the problems."

"The problem is people," Tobey concludes, which concurs with most viewpoints I have heard. "There are too many people using a limited resource. My analogy is, where there used to be room in a section of the harbor for ten scallops and ten people, there are now fifty people, so the scallops lose. And, they're an indicator species, an indication of what's happening in the harbor. We still have a very clean, wonderful harbor, but the scallops have opened our eyes early, and the board is very concerned. And it's much better to do things pro-actively than to try and go back and clean it up, as they have in so many other places."

"This is one of the more active boards of this type?"

"Definitely. Not just SHAB, but Dave has done an amazing amount of work. He's an incredibly hard working person, one of the really unsung heroes of Nantucket in my mind. But SHAB has been very active, and especially in the last ten years in leading the way in many areas. We had a lab functioning, we tried doing a nursery and a hatchery. We've been on the forefront. A lot of people have looked to Nantucket to see what we've done. And we had great help back in the late sixties, early seventies when Bill Klein did a lot of the Zoning Work. He did a fantastic job in terms of preserving the island. And the HDC, despite all the naysayers, has done a fantastic job in terms of preserving the architecture. There have been a lot of groups that have done good work. And I like to think that SHAB has been one of them."

November 19

Sunrise: 6:33 Sunset: 4:18
Temps: Low 31° High 53°
Conditions: Clear, calm

island cup

It is pitch dark at 5:00 p.m. on the Friday before Thanksgiving, and Nantucket is a very quiet island. Many townspeople have already made the pilgrimage to Martha's Vineyard for 'The Game', the longstanding high school football rivalry between the Whalers and the Martha's Vineyard Grapes, to see which team will take home the coveted Island Cup. Nantucket Airlines adds extra flights to accommodate the fans, and provides ground transportation to boot.

Nantucket has appeared in six divisional state championship games, known as Super Bowls, during the 1990's. Their coach, Vito Capizzo, is legendary. He played for Alabama's Bear Bryant, ranks second in the state among living coaches with 250 wins, and the Whalers currently hold the Division Five state championship. He, and the team, have even been written up in *Sports Illustrated.* Be that as it may, the best Nantucket can hope for this year is to spoil the Vineyard's undefeated season. With a mediocre record of five wins against four losses, the Whalers have no hope of defending their title. But if they somehow manage to 'Crush the Grapes' tomorrow, they will at least prevent the

Vineyard from going to the Super Bowl themselves. But that is of secondary importance. A victory means that Nantucket will retain the Island Cup, and crucial bragging rights, for another year.

An entire section of this week's *Inquirer and Mirror* is devoted to 'The Game'. It features quarterback Jon Barone on the cover, and the first article is headlined 'Beating the Vineyard Can Make a Season.' For the record, the Whalers hold a 15-6 edge since Capizzo and then-Vineyard coach John Bachelor kicked in $107.00 to buy the first Island Cup.

Tonight's stillness is heightened by the fact that the winter exodus has begun. People with seasonal jobs are trickling out. Year-rounders with flexible schedules have already departed for holiday visits. The population has dropped from a summer high of over fifty thousand to under ten thousand, on its way down ultimately to a low of roughly seven thousand in the dead of winter.

I visit Marty Morse at Brant Point Marine. He is beginning to put boats away for the winter, covering them in shrink-wrap plastic. Having spent most of his summers on Martha's Vineyard, something he said in an earlier conversation about the nature of the two islands has struck me.

"This island - both these islands. People who live here are very independent. The tourist industry is not something that lends itself to remaining that independent. So an activity that is successful, like fishing, is a great thing. It's not money that's coming in from outside, it's money that's being created here, from photosynthesis. The guy who stopped by earlier today, he's opening now. During the summer he gets clams. Usually he makes a decent living. It's not a big living. He's not gonna become Rockefeller getting quahogs. But, he doesn't have anybody bugging him. He can go out in a fairly small boat, not a lot of equipment. Works when he wants to work - as long as it's low tide. Works at his other things he wants to do. Goes out to the head of the harbor, which still looks quite wild. It's a beautiful place. Sometimes a nice time for the tide is six in the morning, so you're out there at four-thirty getting ready, and you're out ready to go at five. You get to see some beautiful, natural things. Sometimes you might be the only one out there."

I ask Marty what people are buying these days.

"When the season first started the big thing was getting set up - knife, apron, opening gloves," he says. "As the weather gets colder, people are looking for hats." He emphasizes the word 'hats' with a knowing smile. "Waterproof gloves. Warm stuff. We've got some special polypropylene socks, really high performance stuff, originally developed for mountaineering and arctic expeditions. A little bit expensive, but when your feet freeze..."

He leaves the implication to hang in the air: maritime dwellers all know the feeling. "Like, a couple of days ago it was blowing thirty-five to forty, makes the wind chill ten to twenty degrees."

"It was really tough just painting," I agreed. "Can't imagine what it would be like out on the water."

"Well," says Marty, "Just get someone with buckets of salt water and throw 'em in your face."

fishing effort

I ask Marty about the dwindling numbers of scallops being caught each year, with last year's disastrous numbers on a lot of peoples' minds. Marty wonders if any statistics are available comparing, on a year-by-year basis, what is called the 'fishing effort', or number of people fishing, to the size of the catch. In fact, Dave Fronzuto keeps track of both the licenses granted and the yearly total catch, and even has a rough chart of who went fishing on what day. A brief comparison of those statistics with economic conditions on the island yields a pretty common-sense interpretation of the data. The catch declined steadily through the 1980's, spiked up briefly in 1990-1992, and declined again steadily through the 1998-1999 season. But the spike of the early 1990's was at about a third the level of the early 1980's. There was a severe economic recession in the early 1990's, which hit the trades particularly hard. The scallop catch increased because people needed to scallop to make money. But even with an intense fishing effort, only one-third as many scallops were taken out of the water as there were in 1980. And in 1998, there were simply very few scallops to be caught. As Dave had told me in my first talk with him before the season started: 'Every adult animal came out of the water.'

Two scallopers come in to pick up latex opening gloves and other miscellaneous supplies. Working the harbor, they have already gotten their limit and shucked them, but it is now closer to a full day's work instead of half a day's work. These men appear to be in their mid-thirties, on the young side as scallopers go. As Madaket begins to get fished out, more boats have begun to work Nantucket Harbor, where the scallops continue to be very large, but are less densely packed. The younger of the two, a large and muscular man standing about six-foot three, says that he has just discovered that there is no point in fooling around with light dredges. "Just put as much iron on the bottom as you can," he says proudly.

Conversation turns to the rough conditions on Wednesday, which claimed a casualty: Chris Roberts' new boat sank. Waves began washing over the stern, and in a situation like that either the scuppers which allow water to flow out get clogged, or more water washes in than can run out. The added weight of the A-frame makes the boat top-heavy, so it takes longer to right itself, allowing even more water to wash in. A full load of scallops weighs it down even more. And Chris had all eight of his dredges out, which made it hard to turn away from the waves.

Chris was relatively close to shore, and another scalloper named Chris Gordon was fishing next to him, and was able to pull him up near the beach to where he was safe from going down with his boat. He was able to bail out, but spent the afternoon flushing his engine out with fresh water and running Mystery Oil through the carburetor and cylinders.

Sinking is not only a frightening experience, it can be an expensive one. Often a scalloper can get underway again that day or the next, but sometimes the boat and engine incur damage which requires a few days' repair, and that much lost pay as well as the cost of the repairs.

A storm has dragged this boat ashore, despite its mooring.

white knuckle ride

I wonder what Chris Roberts must have been thinking as his new boat was going down. Many Nantucket fishermen have had a similar brush with disaster. Looking back from a distance, the misadventures are often recounted with a certain glow of excitement, sometimes even humor. Ken Holdgate Jr. once broke down off Eel Point during a January storm, freezing, with no way to let his family know where he was. The Coast Guard finally found him near midnight. One man, known to drink frequently, fell overboard near Coatue when his boat hit the Steamship wake. His waders immediately filled with water, anchoring him to the bottom. Luckily, he had taken a deep breath, and started walking toward shore. His head bobbed up before he ran out of air. Another hefty scalloper, weighing close to four-hundred pounds without oilers, fell overboard. Too heavy to be hauled in by either the other scallopers or the Coast Guard, he was towed over to Coatue and beached, where he was able to climb back into his boat. Another scalloper got his foot tangled in the rope and was yanked overboard, trailing the boat and spinning like a pinwheel, until his partner saw what happened and cut the engine.

No one is immune. Even Carl Sjolund had an experience, right at the dock. Attempting to hoist a box of scallops up to the dock, the boat slipped away, and in classic novice sailor fashion, Carl hung momentarily between boat and dock before splashing in. In his words, it took 'quite a long time' to work his way to the surface, finally using the mooring rope to haul himself aboard. He and Charlie Sayle, as teenagers, started sinking once and ran their boat onto a shoal before it sank totally. Another scalloper fell overboard and managed to make it to shore before hypothermia set in. He broke into a house and started a fire to warm up.

Eddie Loring's partner Malcolm nearly got swept overboard. He was bringing the boat across the north side of the island, from Madaket to the harbor, when a sudden squall came up. "All the dredges from one side of his culling board went over into the ocean," said Eddie. "He was lucky, he got pinned to the rail, so he didn't actually go over. Malcolm always carried a machete, often stuck point first in the culling board. The only way to get loose was to cut those dredges off. Always have a knife on your culling board - can be a life saver. For a couple of years there they were talking about one limit per boat. I always thought that was a bad idea, just for safety reasons. There've been so many incidents where it's better to have a second person on board. Just too many things. And we've had plenty of times where we've had to tow people in, and it's always better to have a second person on board."

Behind the gallows humor is the reality that all fishermen live with: fishing is the most dangerous occupation on earth. In 1996, sixteen fishermen per 100,000 died, more than seven times the national industrial average. In Alaska the number was forty-one and a half per 100,000. Entanglement in machinery is a major cause of farm accidents, and most construction fatalities result from falls. Fishermen drown - often without leaving trace or word.

One day Eddie had come to rehearsal, exhausted and bedraggled, and pronounced that he had been sure he was going to drown while scalloping that afternoon. The rest of the cast nodded politely, as though he had said 'the cashier at the A&P shorted me five bucks,' and we proceeded into a three hour rehearsal. I reminded him of it later.

"Nasty weather out there, storms, that's what gave you the biggest rushes," recalled Eddie. "We've been out in thirty-five, forty-five mile-an-hour winds. We'd go out there quite often when the steamship wasn't going. Malcolm had a rule that if we could row to the boat, we were going. It would take us normally about a minute-and-a-half. But I can remember a couple of times where we're literally almost fifteen minutes rowing to the god damn boat! I'd be like 'Mal - you think this is a good idea?' He'd say - 'We're here, we're going.' And I mean, sometimes we'd go out there, make one pass, and we'd realize it's just way too rough."

"We were out there a couple times with maybe one or two other boats - sometimes we'd be the only ones out there. You'd have one hand hanging on the basket, you'd cull with one hand sometimes. Quite often you'd have to hold your dredges on the culling board. Sometimes you'd be hitting chop, you'd hit waves, and your boxes would be falling over and you'd be getting scallops all over on the bottom of your boat. It could physically hurt, too, getting banged around. Especially coming back. To cut through the waves you have to go as fast as you can. If you go slow you're going to get battered about. Those were what you called 'white knuckle rides'. You're holding on for dear life and you're scared enough that your knuckles are turning white. I'd get way up in the bow, looking at Malcolm, and every single wave he'd just be getting slapped right in the face. Totally soaked right through. But we'd still come in with our limit. We'd unload our scallops at the shanty, go right down to the Chicken Box and have a couple of beers. I mean, you'd be shaking. For about an hour-and-a-half you still wouldn't have your land legs. Your whole body's still rocking back and forth."

"How do you keep the boat from filling up with water?"

"Well, on ours we had two pumps. And there were times when I was bailing, literally. Pumps weren't keeping up. When you're towing, you're not moving that fast, and water would be constantly coming in. One of the worst places to go through when it gets like that is the channel at First Point coming back towards town. And the tides usually run pretty strong. It was our last year, and it was blowing pretty good when we first went out, thirty-five, forty miles an hour. We fished in the harbor down towards Third Bend. We'd actually gotten our limit, and we were coming back and got to about Second Bend, and it got really sloppy in there. It must have been gusting about fifty, fifty-five miles an hour. Really bad. We were taking on an incredible amount of water. I mean, the bow of the boat's up, and you're looking at green water coming in. We had both pumps going, and I was bailing as fast as I can, and it was getting higher and higher and higher. And it was, in all the twelve years that I fished with Mal, it was the only time that he ever said, 'Eddie, break out the life jackets, we're going down'."

"It was the only time we ever slapped them on. But for a number of different reasons neither of us were ready to jump. We were knee deep in water, it was literally up to the rails. Malcolm kept his eyes just straight ahead. I could only catch glimpses of him between each wave - and I was getting slapped from behind. I'd always heard that in stressful situations you get an adrenaline rush, you start doing crazy, superhuman things, but I never actually really believed it. That day, I can tell you I never moved so fast in my life. Everything was just a blur. I was lifting three boxes at a time, for moving, for shifting weight around, for balance. I could never do some of those things that I was doing that day. Swear to God, all we had to do was take one more wave, and we were going to go down. Somehow we got into First Bend and the water calmed down. We don't even know how we made it there."

"This is the reason we couldn't give up the ship: basically, the wind was coming out of the northwest, and the tide was running in. We were only about fifteen yards from shore, but if we'd gone overboard, we'd have been sucked the other way. There's no way either of us would have made it to shore. And we probably would have only lasted about three minutes in that water. This was in late November. There was no way we were getting off the boat. If the boat was going down, we were going down with it. Literally, that was the total mentality right there. That's why were working so fast and so crazy on that boat, because that boat was our lifeline. The only way were were going to survive was to keep that boat afloat. We couldn't turn it - we had to keep going directly into the wind. The minute we turned that boat to get into shore, to get onto Coatue, we would have gotten swamped. All we could do was just shoot for a lee. We got one, and it was just in the nick of time."

"Everything looks the same, the water and the sky."

eighty-two years on the water

I am to meet Oscar and Joan Bunting at noon at the mud flats of the H Street landing. I hear the eerie rumble of engines, but see nothing but white and gray. Finally, through thick fog, two shadow-shapes approach side-by-side, as though racing.

November 22

Sunrise: 6:36 Sunset: 4:16
Temps: Low 39° High 52°
Conditions @ 6:53 am: 44°, dense fog, calm
Marine Forecast: Fog, visibility 1 mile or less early, becoming sunny. Wind 10 mph variable. Seas 2'-4'

"Hey, watch where you're going, will ya?!" A voice unattached to a figure. Laughter. It looks like the two shapes are about to ram each other, but with engines roaring louder and closer, and the profiles of two boats growing more distinct with shortening distance, they safely pull up to the flat shore.

"How is it out there?" I yell to Joan. "Can you really see where you're going?"

"No!" she laughs. "If it was up to me we'd still be out there! Everything looks the same, the water and the sky."

She tells me that her husband Oscar navigated by compass, and that three other boats followed them in. Oscar and the driver of the other boat have been bantering all the way to shore. That the fleet is now getting in at noon is an indication that scallops are getting harder to find. Today's fog makes it that much harder, since the scallopers cannot use their visual ranges to find their favorite fishing spots.

This reminded me of another Eddie Loring story:

"I was fishing with Bob Ruley out by Tuckernuck, and the fog came rolling in real thick, and Bob didn't have a compass on board. We ran into Eddie Stojak going in, so we're following right behind him, putt-putting in and putt-putting in, quite a while. So finally we come to a point where we see another boat, we could only see maybe ten yards. Then there's a fourth boat, suddenly then a fifth boat. And everybody's stopping. Next thing all five boats are together and it's like - 'Where are we?' And somebody says - 'Hey wait a minute! I know where we are. This is the Bathtub! We're on the other side of Eel point!' We're like - 'Wait a minute! How come you didn't know where you were going?' 'Well - because I'm following this person.' 'Well, hell I was following this person.' 'Well, hell, I was following you.' And nobody had a compass on them. Turned out Rick Kotalac was the one who started the whole thing. He didn't know he had four other boats behind him. And he led us all into the Bathtub. And that can easily happen, the fog comes rolling in."

Oscar and Joan have come in with six boxes. "There's a lot of small ones out there, but we don't like to fish on them," offers Joan. "Guys come in probably at nine o'clock, but we're not crazy about those scallops. That's why we don't have our limit - could have had it at nine!" she laughs. Oscar mentions that they like to get the very large scallops.

"Six good ones," emphasizes Oscar. "The kind the J.C. serves." He is referring to the Jared Coffin house, a longtime inn and restaurant at the top of Broad Street. "Had some small ones, but we left them there. Like that one there, I'd throw that back." He shows me what looks like a perfectly good scallop, then disdainfully throws it over his shoulder and back in the water. Then he hits me with the question. "You say scollop

Joan and Oscar's daughter, Lynn, plans to visit for her 31st birthday on November 27th. Joan recalls being pregnant with her.

"It was the best year they had had. My husband said 'You're not going scalloping - you're almost ready to have the baby!' So I went to Dr. Voorhees, and he said - 'Oh, if you want to go, go.' So I went."

"You could have literally had her on the boat?"

"We were prepared. We had blankets, all wrapped in plastic. I had this big stomach, and all the boats used to try to help us, which made more work for me. The guys were like - 'Get in! Get in!' I went the morning I had her, that was on a Wednesday, and I had her that evening around seven o'clock. I got out of the hospital Friday, and Monday I was back scalloping."

or you say scallop? You're a Nantucketer if you say scollop."

"I say scollop," I reply, laughing. "That's the first page of this book. Because I noticed how everybody else pronounces it scallop."

"They're from New York."

Oscar and Joan have been fishing together as long as anyone can remember. In fact, Oscar tells me that he started in 1951. He guesses that there were only about three-thousand people living on Nantucket in those days. Joan tells me that Oscar was eighteen at the time, and that she started scalloping fourteen years later, when they were married. Doing my own math, they have eighty-two combined years on the water. Oscar has never done anything but fish, scalloping in the winter, clamming in the summer.

Joan grew up on the island as well, and both her father and uncle scalloped. Her grandmother and one of her uncles was from Poland, but her father was born in New Bedford. As a child living on York Street, she can remember her father and uncle opening scallops down in the cellar.

"I begged to go down, and my father would say, 'No I'll bring you some up'."

"They probably told dirty stories down there."

"Well, no I don't even think it was that way. I think it was ... well, it was men opening, number one. And I was just five years old. It wasn't a place for a little girl to be. So he'd bring the scallops up, let me open them in the sink."

Three trucks back down into the mudflats, creating a virtual traffic jam, which seems bizarre today in this remote, fog-shrouded corner of this rural (at least in the off-season) island. The three boats that followed Oscar in are now offloading, among them the man who had been 'racing' them earlier. His

name is Mickey Callahan, and he is about the same age as Oscar. He has a loud and boisterous laugh and penetrating speech, full volume. Mickey scallops for the first part of the season, then goes to Florida.

"Somebody told me four, five years ago someday I'd wake up and find out I'm not sixty anymore," he explains.

"Yeah, I told him,"says Oscar.

"I been coming here about twenty-five years but I moved here steady about sixteen, seventeen years ago," Mickey says. "I lived on the Cape. Before that I came from Maine."

"So how'd you get here?"

"Fishing. Fishing up in Maine, and then I came down started fishing Woods Hole."

"What do you fish for?"

"Right now, scallops. Before that, all bottom fish. Flounder, codfish, haddock."

"We both used to go clamming," says Oscar. "We're the last of the Mohicans."

"Yeah I know it! All of a sudden we look around - we're the young guys there. We're the only two. Young kids say, 'You guys making money here?' Say - 'Get out there and see what it's like.'"

"I don't think anybody should go broke on Nantucket," Joan adds emphatically. "There's always a job to do."

"That's true!" nods Mickey. Now the three of them get on a roll, overlapping each other, laughing boisterously at each comment.

"You know when I hear people say that they don't have any money - and they got two legs and two arms - it's their own fault."

"Yep."

"Because there's many jobs you could work. I've worked three jobs in my lifetime."

"I work two now!"

"Yeah, I do too!"

"I still cut grass in the afternoon and go scalloping in the morning."

"They give us so much money for Social Security - so actually we shouldn't have to work. Hah hah!" laughs Oscar, stretching out the two syllables for effect. "Pays for the gas to come out here!"

"Wait for Social Security and there's none there."

"These are our golden years, you know!"

"Yeah, the only thing golden about my years is my urine, for god sakes!" shouts Mickey.

Though it sounds on paper like it might be a scene from *Grumpy Old Men,* in actuality these three are still laughing about the whole situation. In a moment of reflection, as a cigarette is lit, Mickey asks:

"Oscar, what would you do if you couldn't work?"

"What would I do? Get a job. I'd go work for somebody else."

"Me too. Like my daughter says: 'Why do you work, Dad?' Don't want to die, for Chrissakes, that's why I work."

"So was it hard work out there today, Oscar?" I ask.

"Every day is a struggle," he says as though martyred, but fools no one. "We had some good fishing there the first three weeks. You know, there's one thing worse than working - that's not working. I enjoy work. Actually, this is fun."

"Right!" says Mickey. "You enjoy it. It's something you do and enjoy."

"When they come good," qualifies Oscar. "But when they don't come too good, then it's work."

"Today, we're telling you all this stuff," says Joan. "Today was beautiful. But I can remember when the sleet would be down in my face, and tears would be rolling down. My toes were cold, my hands were cold. But you have to make the money to pay for, you know, shoes for the children. But today it's a lot easier, because they're all grown up."

Oscar's grandfather, Oscar Folger, operated the freight and mail service in 'Sconset, and had the first trucks on the island. "They didn't allow the cars to go downtown," explains Oscar. "Had to stop at the rotary. He'd drive the truck in from 'Sconset to the rotary, unload it, put it on the team of horses, go down to the boat, go back to the rotary, unload it, and put it on the truck. But then he finally smartened up. Instead of unloading the truck, he just hooked the team of horses right to the truck, and towed the truck downtown."

"So your mother was a Folger, I gather. And that goes all the way back the original Folgers. So you're related to Ben Franklin."

"That's right."

"But scalloping wasn't a family thing?"

"No, just my father," says Oscar. "My father lost his life going scalloping. He left Nantucket harbor, and he was coming up here to the west end to go scalloping, and I guess it was a terrible storm when he left. And for some reason he didn't make it."

"How old were you?"

"I don't know - about four or five."

"I think you were almost eight," says Joan. "Almost seven or eight."

"You were just a kid? And that was ... probably during the war? That must have been rough on the family. And when you grew up, you went scalloping anyway?"

"Yeah. It's a good living, you know. It's a lot of fun, lot of work. If you're willing to work, you can make a good living at it. Back in those days there wasn't anything else to do. You know, there wasn't houses being built like there are today, so this was the living. You depended on it."

As much as three old pals can enjoy fishing together on a warm and foggy November morning, it has always been a dangerous occupation, and the currents, tides, shoals, and weather conditions around Nantucket are notoriously treacherous. Over two thousand shipwrecks have occurred in these waters, and countless lives have been lost. Three years after Oscar's father died, Joe Viera Jr., Eddie Stojak's step-brother, went out scalloping near Fourth Bend, toward the Head of the Harbor. A snowstorm had blown in, but in those pre-building boom days, there wasn't much other work for a scalloper who needed to pay the bills, so he went out. Joe didn't return, but the cabin of his boat was found on Quaise. Two months later his body was discovered on Coatue. Most people theorized that his boat must have exploded. He was not the first Nantucket scalloper, nor would he be the last, to lose his life on the water.

Many of us have had a brush with accidental death on the highway, but a screech of tires, a Doppler-distorted horn, and the incident is over before our heart rate even picks up. Or a debilitating illness slowly but inevitably brings helplessness, then death. Not many of us have, like Eddie Loring and Malcolm Barreiros, found ourselves engaged for fifteen or twenty minutes, maybe half-an-hour, in a constant and real battle against death - where all human effort may or may not save one's life, but giving up assures death. How quickly or slowly did Joe Viera meet his fate? How long did Oscar's father struggle before succumbing to that storm half a century ago? What thoughts were going through their heads?

Thanksgiving

Sunrise: 6:39 Sunset: 4:14
Temps: Low 45° High 55°
Conditions: Mostly cloudy, misting

"One of the things I loved about scalloping was just getting up early in the morning, getting out there on the water. Especially when it's really calm, really quiet."

gentleman tow

Eddie Loring had also told me of a Thanksgiving tradition. In Eddie fashion, he set the scene beforehand:

"One of the things I loved about scalloping was just getting up early in the morning, getting out there on the water. Especially when it's really calm, really quiet. It's a great way to wake up for the day. It wasn't all intimidation and all threatening, always being on your guard. Beginning of the season is pretty mellow - it's when the pickings get harder and harder people start guarding what they can get. But the camaraderie was always great, and it's always fun when you're out there. You're working, you see somebody you know, people will come over they'll talk to you, or you'll go over and talk to them. Lot of times you'd be culling off and they'll come over and cull off next to you. There'd be a bunch of us, we'd take lunch breaks together, tie our boats together at lunch time. We'd drink and eat, shoot the shit for a little bit. Later on in the season you'll see a lot of boats tied up together."

"We had this thing we did for years. Thanksgiving Day, bunch of us would get together, sometimes here in the harbor, sometimes out in Madaket. You brought your six-packs out there, your lunches, and everybody would start at the same time, same spot. Threw all their dredges, same speed. Make a clean sweep of an area. Come around, pull up together, and pull all the dredges in together. If you got your dredges in before another boat, you'd wait until that boat got all their dredges in. And everybody together, same thing again. Clean sweep. Sometimes we'd be so close together doing this, you'd see five or six boats together. You could walk from boat to boat to boat while you're fishing. Smitty had Moonshine out there, his golden retriever. Moonshine was trained to carry beer in a paper bag. Would go from boat to boat to boat with a paper bag with beer in it. People would grab a beer out of the bag and the dog would walk back across, culling board to culling board. It was just a lot of fun. It was called the 'Gentleman Tow'. That was a tradition for a number of years. It would usually be me and Malcolm. Smitty would be with us. Sometimes Pennel Ames, sometimes Andy Roberts. Al Sherburne. Quite a few of us regulars. The next day everybody'd be off in a different spot doing their own thing again."

Meanwhile, the Grapes sank the Whalers last Saturday. Accursed Martha's Vineyard now holds the Island Cup and the right to flaunt it for the next twelve months. This is not a situation the community swallows easily. Nantucket has lost the Cup to the rival island only six times in twenty-two years. So this Thanksgiving, instead of travelling to the mainland to cheer the Whalers on in the Super Bowl, people will bemoan the season's finish, watch college football on television, visit neighbors, and catch up on news and gossip. Among the most interesting bits of news: Allen 'Froggy' Renaud, who used to be my mailman, won $4,000,000 in the lottery. As they say in New York, 'Hey, You Never Know.'

"Lot of times you'd be culling off and they'll come over and cull off next to you. There'd be a bunch of us, we'd take lunch breaks together, tie our boats together at lunch time. We'd drink and eat, shoot the shit for a little bit. Later on in the season you'll see a lot of boats tied up together."

November 29

Sunrise: 6:44 Sunset: 4:13
Temps: Low 38° High 46°
Conditions: Clear

hunting season

A ten day deer hunting season began today. Hunters flock to Nantucket from all over New England with shotguns and hunting licenses in hand. 111 deer (67 bucks and 44 does) were bagged on opening day. If such statistics are kept, Nantucket must certainly rank near the top in ratio of deer killed per acre in the country. And that is only counting car collisions. If you add the hunters, there would be no contest.

Deer hunting on Nantucket is not really a time-honored tradition, as 350 year-old towns measure these things. In fact, the first recorded deer hunt took place quite recently, in 1935. During that initial hunt, Art Stojak's father was shot twice, in what the newspaper suspiciously reported as a 'hunting accident'. He survived, and the culprit with extraordinarily poor aim later surrendered himself to the police.

There is a certain amount of ambivalence on Nantucket regarding hunting. As in most rural communities, hunting is a way of life, and not as far removed from the practice of earning a livelihood as it is in more urban communities. A farmer or rancher or fisherman lives and provides life to his or her fellow man by wresting living things from the ground or the sea. Wildlife can be a constant threat, a source of food, or both. Hunting them is a natural and obvious part of the job and the life.

But vacationers and people who love nature find themselves attracted to Nantucket and places like it precisely because of their rural feel, and along with that love of nature often comes a similar love of the animal life within nature, hence often a repulsion toward hunting or even fishing. Not to mention the genuinely disconcerting sound of shotguns at almost any locale during this ten day period, and enough stories of dogs or picture windows being shot, combined with nightly images of bars full of sloppy drunk hunters, to give any reasonable person cause to fear for their safety.

The wild cards in this witch's brew of conflicted emotion are deer ticks and the highest per capita incidence of Lyme disease in the world. Virtually everyone recognizes that with no predators, the deer herd on Nantucket would soon grow so large that starvation would be rampant. People also recognize that with the growth of the deer herd comes the growth of the deer tick population, which transmits the debilitating and sometimes fatal disease. So hunting retains an important position in the scheme of island life.

Naturally, Eddie Loring had some stories on this too, which he had recounted on spaghetti night, toward the end of our supply of beer and wine.

"Hunting out on the water was always a great thrill. Smitty does it, I don't know how many people do it anymore. Keep a shotgun on the boat. Soon as duck season opened, before we'd throw our first dredge, we'd try to bag our three birds. Sometimes we'd even offer the wardens one, if we'd caught a goose or a brant. They'd love to see us coming in because we'd have our limit, but we'd also have three birds hanging from our mast. That was a great sight! We had a ritual out there that a lot of people would find disgusting, but you would do this with a brant. Brant is not a goose and not a duck, but it's in the goose family. That's how they got Brant Point: there was tons and tons and tons of brants. When we gutted out the birds, we'd always take the heart. Particularly if it was still pumping, because a lot of the times the birds wouldn't fall dead. Our ritual basically was to take the heart and actually eat it right there while it was still pumping, it'd still be warm. That got started on a bet and a dare, and then it turned out we'd be doing it for years."

"Was it small, like one bite?"

"No, actually it was good size."

"You have to cut it up?"

"Well, you could cut it up, or you just take a bite, give a bite to another person. Malcolm would cut out the heart and he'd say yell over to Smitty or Pennel or somebody - 'Hey, you guys want to come on over and have a bite?' Most of the time it'd be no, but once in a while somebody would. Some people were very surprised, because actually it was very tasty."

"Does sound disgusting."

"It would make us a little grisly and a little more intimidating out there. You know, you're out there eating bird hearts, and you got blood all over your beard, we'd start shooting out there and people would get very nervous. You kind of want to stay away. Never shot a deer out on the water. In fact, being the great humanitarians that we were, there were a couple of times we'd see deer swimming across the harbor, and the shoreline was just lined with hunters. Twice it happened right around Folger's marsh. They'd chased the deer in there and they're waiting for it to get tired and come back. Well, first time this happened I made a lasso, slipped it right over the deer's head, and basically steered it. It would swim right alongside the boat. Ran right over to Coatue and let it go. Now you've got all these pissed-off hunters on the other side. We had a great laugh over that."

"One year on our way out to Madaket we hit a deer on the Madaket road. Got outside, picked it up, threw it in the back of the truck. Got down to Fisher's landing, went to get our scallop gear out, looked in the back of the truck, and the deer was standing on its front legs looking at us. Now, unfortunately, its back legs are broken. And it's probably the only time Malcolm said he ever felt really bad about killing an animal. I had to hold its antlers and Mal had to take a hammer. So we brought it down to the boat, put it on the culling board, and everybody that day was out fishing around Pirate's Cove. Malcolm started gutting the deer on the culling board, and the water around our boat just started turning bright red. People were going by saying 'Jesus Christ, Malcolm! What the hell kind of scallops are you getting?'"

"That's part of the thing about the camaraderie, too, that was just plain fun out there. Like one of our sports was we'd be fishing alongside one another and hauling all kinds of stuff off the bottom. Lot of times you'd come up with a toadfish or small monkfish or things like that. Anything really odd or unusual or anything that looks, you know, really awful. Conch that's dead or three-quarters dead, real slimy. You see your buddy over there and you'd be like 'Hey! Hey Smitty!' And you'd wing this awful looking thing over into their boat. You'd always get a chuckle out of that. And they'd do the same to you - you'd be pulling in your lines and all of a sudden you'd hear this 'thud' and there's this big monkfish just sitting there on your culling board. Or it'd be slapping around the bottom of your boat, something like that. We had an unusual experience once - in our last year. It was getting late in the year and we went to the boat early one morning and I started straightening up in the bottom of it and came across this sac. Thought - 'Jesus Christ, what the f--- is this?' I pick it up. Malcolm looks at it, I'm looking at it - all of a sudden we realize - it's a deer nut sac. It's a pair of deer balls. Our buddy, our good friend Smitty, who had been hunting and who was bragging about how he'd just copped a buck the day before, thought it was a big practical joke to toss the balls onto our boat. My immediate reaction was to throw it over, but Malcolm says - 'No no, save it. We're going to see Smitty again'. So we get out there in the middle of the harbor, and we haven't seen Smitty. He's still in bed, he hasn't gone out fishing yet. We'd thrown our buoy and we're making our tow, and this fellow whose name I won't mention came up and jumped on our tow. First off, Malcolm got very irate about it. And suddenly he says - 'Okay, he obviously needs to get a clear message here.' Well, he'd thrown a buoy at the opposite end. So we pulled in our dredges and he had all his dredges out. And we putted up and pulled his buoy. Whereupon we went and tied the deer nuts to his buoy. And then turned around and threw his buoy back in like a couple hundred yards up harbor."

"Did the guy see it?"

"Oh, he knew we pulled his buoy. He was fuming that we had pulled his buoy. He was really pissed off about that. We were just laughing and laughing, and he goes over and pulls up his buoy - and the minute we saw him pulling up his buoy we started laughing even harder, because we knew there was a sac of deer nuts on there. And that got him - that set him off real good. He was swearing and cussing at us - finally he pulled out a hatchet and slammed it into his own culling board. We probably would have taken it seriously except we were just laughing so hard about the deer nuts that were clinging to his buoy. I don't know what you're going to make of that story."

"I'd love to know who the guy was."

"Well, I'm not going to tell you 'cause he'll still get angry to this day. We had our friends and we had our enemies out there, a bit of both. Mal was real wild-looking. He had long black hair with bits of gray in it, came down way past his shoulders. Scraggly beard and a long, long mustache. He was part Portuguese, part Irish, real physically built, real strong. By looking at him, you just did not want to tangle with him. Basically, he looked like Blackbeard the Pirate. He was a wild man - he had a pretty good temper, and it was very intimidating. There was a couple of years there we'd take the machete aboard the boat with us. Basically we used it for cutting lines - sometimes you'd have to. You got in trouble out there and sometimes you'd have to lose a dredge or two. The best way was a sharp knife, and the machete worked perfect. But the machete also kept a few boats away from our tows, too. Malcolm wasn't afraid to brandish it. I remember once in a while he'd threaten to cut somebody's line with it. He'd be out there swinging that thing around his head, yelling and cursing. You've probably seen Malcolm's boat before. It was actually very famous. Because the name of his boat, written in great big red letters, was *Blow Me.* Everybody knew the *Blow Me.* For years people coming in on the Steamship Authority would see this scallop boat out there that says *Blow Me* on it. And our little dinghy that we used to row out to the boat was called the *Blow Me II.* And he used to tell his kids that it was the 'Blow Me Down', only the 'Down' was below the waterline."

"Well, that adds even more interest to what you said earlier. What kind of an idiot would jump on the tow of a guy who has three dead ducks hanging on the mast, looks like Blackbeard, swings a machete, totes a shotgun, eats bird hearts, and has a boat named *Blow Me*?"

A 'pillow' at Evie's shanty is becoming a rarity. Now that December has arrived, scallopers are struggling to bring in a load of 'beans'.

December 1

Sunrise: 6:46 Sunset: 4:12
Temps: Low 32° High 37°
Conditions: Blustery, gusts to 50 mph

beans

The second month of the five month scalloping season blew in hard, with temperatures in the low 30's, winds gusting up to around 50 miles per hour, and the wind chill factor at three degrees. Kenny Aguiar gave it up today. It was a good month of November for him though - he made his Christmas money. Kenny, with whom I have often shared a breakfast counter at the Downy Flake, started the season with his leg in a cast. He tells me it used to be you could make real money, but now it's Christmas money. He began scalloping at fourteen, and his family, on the Dunham side, has been on Nantucket since 1725.

Over at Evie's shanty, the conversation is cheerily laced with disdain at the load of 'beans' that has been brought in by one of their regular customers who is still fishing Madaket. The fact that beans are beginning to show up is the first sign the harvest is thinning. And the first stirrings of a labor revolt. Since openers get paid by the pound, generally twenty percent of dock price, the math is obvious: a hundred 'beans' per pound of scallops versus, say, sixty-five 'marshmallows' equals a third more work for the same pay. And if beans are going to become the norm, renegotiation will quickly follow. The price has climbed back to $11.00, up from the $10.50 it had been for a while. More fishermen have moved into the harbor, and the in-town scallops are large this year, but the question remains as to how many there are.

Evie

"We fished one year, twenty-nine years ago, up at Wauwinet," says Evie,

"And the scallops, the inside of the meat, were as big as half-dollars. They were huge. It was just incredible the size of the scallops. I've never seen them quite that large before. Although, these are close." She holds one up from an in-town fisherman which is, indeed, very large.

I had heard some tales, going back forty or fifty years, that they were always big like that.

"I don't know," says Evie. "I know when the Haulover opened, for some reason it just flushed the harbor out. They rejuvenate themselves or something."

"It draws the tide out faster," says Evie's brother, Hank Garnett, who has just stepped in. "Instead of taking twelve days to empty the harbor, it probably takes two days, so it flushes all the pollutants out of the harbor big time faster." Hank and Evie have scalloped together often in past years, and I ask how long he plans to keep it up this season.

"I don't know if I'm going to go all year. I might go back to carpentry. Depending on what the price goes to. Last year I finished selling at almost fifteen dollars a pound."

"But you were all day long and what, like two boxes a day?" says Evie.

"Two boxes, three boxes a day. I don't know, you tell me. We went out and got four-and-a-half boxes."

"Yeah, but that was after a good blow." Evie tells me that after a storm, a lot of scallopers will go out. "They all come out of the woodwork, because it takes all the garbage and pulls it onto the shore, so you don't have to cull through all the shack." I ask her how late into the season people might fish this year.

"They'll go until it's too hard to go. Until it's not worth it. What happens is that you tear your motor up. You're pushing so hard, driving through so much shack, and hauling it up over and over again. And if there's seed out there, you don't want to take that, because then you won't have any scallops next year. And it's not worth it to burn up your motor and spend that many hours out there for two boxes, three boxes."

Evie asks me if I had heard about Chris Roberts' boat sinking, and how he was lucky that he was in shallow water and that other scallopers were around. She recalls a stormy morning a number of years earlier, not long after she was first married.

"Edwin Nickerson, he lived right behind us on Union Street. His mother and father lived in the front house, and they let Edwin build a house in the back. He had an old jalopy and I could hear him leave every morning, because his truck needed a muffler, and it would go vmmvmmvmmvm-mvmm. So I said to Elliot that morning - 'You going scalloping?' He says

'No, I'm not going out in this.' 'Why is Edwin going out?' I asked. 'I don't know.' But he had fished a long time. Nice, nice man. Later that day, they said that Madaket Millie had found him. He probably was checking his boat to make sure the lines were okay. They row out in the little dinghies. Either the dinghy flipped over, or he lost his footing, or - you know. Whatever happened ... he drowned. And of course you've got all those clothes on. You know, boots fill up with water. I still can remember to this day."

"How old a man was he?"

"He was young. I think he was in his forties."

The click-clack, click-clack of scallops opening, multiplied by four people working the shanty, fills the silence. I remember that a young scalloper had drowned, a few years back, when the dredge lines had gotten wrapped around his foot and he had been pulled overboard. Evie and Irene tell me this was Frankie Day's brother. I ask if they had been fishing together that day.

"No," they both say in unison and with meaning. "He was fishing alone," says Evie. "That's one of the things that the guys have to be careful about. You've got all those ropes and stuff, that you don't get tangled up."

Elliot breaks the mood up by sliding a pile of scallops over toward his mother and grandmother, who protest that they don't want anymore. But they are closing in on it, the piles are getting manageable. Another man, grizzled and weathered, opens the door and lets a cold blast of air in. He has come to pick up his scallops. Evie introduces him to me as Jim Crecca. I mention that I have heard his name quite a bit.

"Uh, oh," he says. "That's bad."

"No, that's good!" laughs Evie.

"You must have been talking to all the wrong people."

"Everybody says Jim Crecca's one of the best," I say, which is true.

"They got that wrong."

"That's what they say. I'm writing a book called 'Scallop Season'."

"A horror story?"

Evie laughs harder at this. Somewhat modestly, Jim takes his scallops and slips out.

"They work hard, let me tell you," says Evie. "They're hard working people."

DIVING SYSTEMS CE 0403 MAX 35 BAR/508

December 16

Sunrise: 6:59 Sunset: 4:12
Temps: Low 41° High 52°
Conditions @ 7:53 am: 49°, mostly cloudy, wind ESE 6mph
Forecast: Cloudy and breezy, becoming sunny and windy.
Marine Forecast: Gale Warning. SW wind 15-20 kts increasing to 20-30 kts. Seas 4'-8'

diving for them

You can count on two hands the year-round businesses in downtown Nantucket which have remained unchanged since the 1970's: Brock's and Congdon & Coleman, the two side-by-side insurance companies; Nantucket Pharmacy and Congdon's Pharmacy, the two side-by-side pharmacies featuring old-fashioned soda fountains; the Brotherhood restaurant; The Jared Coffin House Inn; Mitchell's Book Corner; the Gaslight, a one-hundred seat movie theatre; the Nobby Shop; Nantucket Looms; Pacific National Bank. Most have modernized to some degree, but The Sunken Ship hasn't replaced a shingle in that time. The floors are still bare wood, sanded smooth by years of summer wear. They carry an odd assortment of resort and recreation gear ranging from kitsch to fashionable to useful. But above all, as the name and the full U. S. Navy dive suit, reminiscent of an old horror movie or a Jules Verne novel, in the doorway suggest, it is a dive shop. Its owner, Phil Osley, dives, and he runs The Sunken Ship because he dives. He teaches diving, he dives commercially, and he dives for scallops. A drunk customer enters, claims he knows Phil from years back, and gladhands everyone in the place. Although Phil's expression indicates he does not really recognize the man, he is nonetheless patient with him. I have heard that patience is not a characteristic Phil is known for. He is known for being vociferous and outspoken, especially if it concerns the harbor and scallops.

This year only he and a man named Pete Kaizer are diving commercially. They are still getting their five box limits, but scallops are beginning to thin out, and the price has dropped to ten dollars. Pete had to work three different locations today in order to catch his limit. I ask Phil how his own day went. He has just changed out of his diving suit and looks exhausted. I quickly understand how he got his reputation for being outspoken.

Pete Kaizer and Phil Osley are the only two commercial fishermen diving for scallops this year. Here, Pete prepares to wade out to his boat at the H Street Landing. The red flag with white diagonal stripe on the back of his boat alerts other scallopers that a diver is in the water.

"I started out late because we were working in a mudhole. The dredgers are in there - they come down to our spot because they can't find their own spots anymore. So they follow the divers." He is more resigned than upset at this, getting in a semi-good natured dig. "Of course, they're right on us, because they know if the diver's got his boat anchored, there's got to be scallops there. So they see where we are, and the next day they're on us. They tear it up so you can't see."

Phil started scalloping shortly after he arrived.

"In those days, in the early seventies, you could throw your anchor anywhere there'd be scallops," he says. "Not anymore. There weren't that many people going. Price was only a dollar sixty-five. Openers were thirty-five cents a pound. But you could come out and make over a hundred dollars. It's harder diving for scallops. A dredgers got eight dredges, right? Like towing eight divers behind his boat."

"So you are one of the few people that has seen the bottom of the harbor around here for thirty years. What have you seen?"

"Oh, I've seen what's going on. There's only one thing. Pollution is an issue they say. Lately the biggest issue has been taking the nub scallops, the tiny, tiny growth rings. That will spat again. We've been fighting with the town and the Board of Selectmen, the powers that be, to get them to enforce the law, which says you have to have a well-defined growth ring. Because for the last couple of years, the well-defined growth ring kept getting smaller and smaller and smaller. So any scallop out there now has a well-defined growth ring. Some dredgers would say, 'Do you have a microscope? I'll show you the well-defined growth ring.' That kind of attitude is what will kill this fishery."

"You'd say that's probably the reason that scallops are so slim?"

"A major, major reason. Taking the babies. An example: we did tests. Took those nub scallops, took them to the lab. One biologist said most of them are going to die. He was a biologist who worked for the town, so everybody listened to him. But the other biologist said, 'No, they're going to live'. And as it turned out, he was right. Over ninety percent lived, and had a spat again. So now, Dave Fronzuto, who saw that there was a lot of seed and a lot of those nub scallops, shut down Madaket last year. This year, there's a bumper crop. Those nub scallops now have a second year growth ring. There's no disputing that evidence. They say you've got to do more tests. Naw naw naw naw naw." Phil shakes his head in emphatic disapproval, as a coach might to a little-leaguer trying to convince him that the stupid play he just made was actually a smart one. "You don't have to do more tests. It's pretty clear. Nature tries to that says 'a well-defined raised growth ring'. And then he goes - 'Well, it is a well-defined growth ring.' That little thing!?" Phil says, virtually chuckling, he is so dripping with derision. "That letter explains what a well-defined growth ring is. That Fronzuto, I give him credit. He went out on a limb to do that. He's got a lot of people that don't like it, because of this attitude. But all the family scallopers and all the commercial guys that were getting their ten boxes in an hour out in Madaket are saying - 'Hey, maybe there is something to this.' And a lot of the people that don't go any more, and a lot of the old-timers, like George Andrews, who's sick now. Too bad you can't go talk to him, but he's in bed. He's really hurting. But last year he's at Hussey Shoal - dumping off the nub scallops and the seed, because they never took them. And the guy behind him is dredging, picking them up. Jeez. They just keep taking those little babies, not realizing that one of those scallops could make a hundred thousand."

go see george

I stop in again at Carl Sjolund's shanty. He and about a dozen guys are still fishing Madaket, but Bruce Cowan has moved into town. I ask who else has moved into town.

"I don't know. Lots of guys went to town. Hard to keep track of everything. You go, and you come home, you have lunch, then you come in here. You know. Go down and sell 'em, you get home at five o'clock."

This particular conversation is one that reminds me of what Dave Fronzuto had said on the second day, and what a lot of islanders already know. You work at one season until you get exhausted, you have some slack time while gearing up for another season, and then you move into that other season, usually at a different kind of job. Although Carl is as energetic today as he was on the first day, for him to even grudgingly acknowledge that he has worked a long day is a sign the season is wearing on him.

On the one hand, finishing at five o'clock may not seem too tough to many. But when you have started at five or five-thirty a.m., spent six hours out on the water with a temperature between thirty and forty degrees, and with a wind chill factor in the teens, and the work is all physical, on your feet the entire time - it takes a toll on the body. Most of the fleet lasted the first three weeks. Another bunch lasted the next three weeks. It will be tough going from here on out. Carl tells me that one of the men still fishing Madaket is Eddie Stojak.

"You and he have been scalloping about the same number of years, haven't you?" I ask.

"Oh, he's been at it longer than I have." I mention that everybody has told me that Herky Stojak is the King of the Fleet.

"Yeah," says Carl. "He's the best scalloper that ever was."

"That *ever* was?"

"That I know of. Goes right at it." Knowing Carl Sjolund as I've come to in the last few weeks, this is simply the pinnacle of praise that can be bestowed upon an individual.

"You talk to George Andrews yet?" asks Carl. He had mentioned once before that I should talk to George Andrews, and I somewhat sheepishly admit I haven't gotten around to it, though it is now the second time today I have been admonished to look him up. "You've got to talk to him. He's eighty-something years old now. I don't think he went at all this year. I saw him the other day with a cane. I think he's at least eighty-two. He had a big catboat that his father had had built, up to just a few years ago. When I was a kid, I went scalloping with him on Saturdays. Lot of the old catboats had Model A engines in them. He'd take the engine out of the old Model A, take the Model A to the dump, save the engine, and put in in the catboat. He always had a spare one right underneath the bench. Very simple - George would make it work."

Carl has said this last phrase with unabashed admiration, thinking back with clarity perhaps forty years to George tinkering with those engines, which themselves had to be already forty years old. I note that people seemed to have been very resourceful back in those days.

"Well, even in my time it was really a hand-to-mouth operation for existence out here. You know, you didn't have all this crazy boom in real estate, people begging you to work, stuff like that. God, it's unbelievable. No lack of work for anybody that's on the ball, anybody that wants to work, has half a brain. Like I always say, all you need is to be honest and show up."

I tell Carl about talking to Oscar and Joan Bunting, and how she had had her baby on a Wednesday and was back scalloping on Monday.

"I remember that! The guys would come by and offer them some scallops so she'd go the hell home!" he laughs. "Had everybody a nervous wreck!

I think that was a big year, too. I think that was about thirty years ago. Had a season in Madaket that was just unbelievable. The whole area's changed so much out there. Tuckernuck was a huge area. A lot of boats would go to Tuckernuck every day, no matter what. No matter how thick the scallops were in Madaket, they'd go to Tuckernuck. I'd say a third of the boats would go to Tuckernuck every day, that much less pounding Madaket got."

"Anybody go out there any more?"

"I saw a couple boats out there today anyhow."

"That's a pretty long trip right?"

"Oh, it don't take 'em long to get out there with them outboards. It's all shoal water, you got to be careful getting in and out. You've got to know how to twist and turn your way in there. In the morning, at low tide, when it's dark, you want to know just where you're going. You don't want the tide to fall and you put her aground. You run aground, you're waiting for the tide to come in, seems like you're waiting forever."

"Currents pretty tough out there?"

"Not right where they scallop, but if you're handy to the opening, the tide runs pretty strong there. Nothing superdangerous, but if you did break down there, you'd anchor up, somebody'll see you. You know the guys, they're all pretty good about keeping an eye on everybody. The Coast Guard never comes out there, so when we're busted, we take care of it ourselves. And we've all been through it, you know. Everybody's been on the other end of the towline at one time or another. It's all part of the business. It's all the rocks and shoals of the scallop business."

As Carl shucks his catch and night approaches, conversation drifts off into a range of topics, from Chris Roberts' boat sinking, to working a crab boat in Alaska with Charlie Sayle, to the *Exxon Valdez* and its effect on the small fishermen in Prince William Sound. I feel I've just received an intensive course in fishing. But modest to the end, even as I leave, Carl says:

"I don't know if you learned anything."

"No, I did!" I protest.

"Did you? Well it probably wasn't here. But you go down and see George Andrews."

December 18

Sunrise: 7:01 Sunset: 4:13
Temps: Low 25° High 41°
Conditions: Slight overcast

the fish house

Everybody should get to live on Easy Street once in their life. My turn came in the spring of 1983.

Easy Street fronts that familiar little sliver of water between Steamboat Wharf and Old North Wharf known as the Easy Street basin. The Skipper restaurant with its showtune singers was still there, undulating on its pilings out over the water. So was the old Steamship Authority terminal, which was not much more than a faded industrial-green wooden shack where you bought your tickets, and an open platform where you waited for the boat. Sheaflike stacks of creosote pillars groaned and creaked against the weight of the docking ferry. And only a slim dirt embankment separated car from harbor, a topographical arrangement that once resulted in my dog Chula hurtling out of a Volkswagen van directly into the drink. Good thing for her it was high tide.

I took many photographs from that spot, but my favorite is of a ramshackle old building sinking into the water. The evening sun bounces crazily off the big 12-over-12 window on the second floor, the panes askew with age, bereft of putty. The weathered door, unpainted since installation, is gray and streaked with rust from the cast iron strap hinges. The pilings are mismatched and oddly angled, supporting what looks like a porch but is really a dock.

I have kept that photograph on my wall in every place I have lived. I look at it several times a day, and I have never grown tired of it. Today, that building still stands, unchanged, and I have just met the man who owns it. His name is George Andrews, and he is eighty-three years old. He has been fishing for sixty-six years - dredging, dragging, netting, potting, dorying, tonging, raking, handlining, longlining, you name it. But his scallop boat rests quietly in the Easy Street basin this year. For the first time since the Depression, George Andrews is not on the water around Nantucket.

George is recovering from pneumonia. He acquired it in October, cleaning the pigeons out of his 'fish house' as he calls it, getting ready for his sixty-seventh season. And George has fished longer than he claims - he only counts his years as a commercial fisherman. As a child, while the Twenties were roaring with Prohibition and flappers, gangsters and Ziegfeld girls, George was culling scallops for his father James aboard his catboat *Wonoma.* While the market was crashing, George and his brother Clinton were learning to mend nets, bait hooks, coil lines, and find fish. By the end of the Depression he was making his own young man's way in the trade. And by the second World War, he was making a solid living at it.

Through the Depression, World War II, the McCarthy era, the Eisenhower years, and the Camelot Days of the Kennedy Administration, George hauled fish aboard the *Wonoma*, which he had taken over from his father. Props became jets, jets became rockets, and George fished. Wireless became television became internet, and George fished. Steam gave way to combustion, combustion gave way to fission, and George Andrews fished. He plied his trade in pretty much the same way half-shareman Richard Gardner did to catch the cod that paid the proprietors' taxes.

When asked about his future, George applies unsentimental judgement, imbued through decades of summing up his daily endeavor in simple terms: how many pounds of fish in his hold. He is eighty-three years old and recovering from pneumonia. He knows the strength and stamina required to take fish out of the sea. When the new millennium arrives, he doesn't figure to be out on his boat. My sense is that if the collective will of the fleet could somehow transmit enough strength to George to get him out on the water alongside them for one last tow, it would happen.

George lives in a whaling captain's house on York Street, built around 1830 by an Andrews forbear. We sit in a living room with sun streaming in, next to a small stand filled with nautical books. I see one about knots, and one about catboats. I am curious about the house down on the Old North Wharf, off Easy Street. George talks in a weakened, hoarse voice, but he does not truly appear to struggle, nor does his delivery slow down because of it.

"My father bought that. That was always a fish house. Even before he got it, that was mostly what it was used for. It was built right after the fire. The big fire, you know?" George says this in case I am not familiar with the Great Fire of 1846, which was the first event in a run of bad luck that decimated Nantucket's fortunes until well into this century. "There wasn't any lumber yard then. What you needed all came on one vessel. There are very heavy timbers in there, so they put the lumber upstairs. After that a fishing company got it."

"What exactly do you mean by 'fish house'?"

"Well, it's full of nets. And we opened the scallops in there. We cleaned the fish. Used it for storage. All kinds of rakes. We used to get clams."

I ask George to tell me about fishing.

"1933, that's when I started. 'Course I went before that, with my father."

James Andrews had had the catboat *Wonoma* built in Osterville in 1902. A large fleet of the distinctive wide and shallow vessels graced the harbor in those times. Their width gave them a very stable platform, they were maneuverable, could navigate shallow water, and even under sail power, were able to pull dredges. The *Wonoma* was named after an Indian princess, the daughter of Wauwinet, the most important sachem on Nantucket at the time of the settlers' arrival. Legend holds that Wonoma possessed a gift of healing, and had cured a neighboring tribe's mysterious disease. She had also secretly fallen in love with their young sachem, Autopscot. When a later dispute threatened war, she risked her father's wrath to warn Autopscot of an impending ambush, and the issue was resolved peacefully. Wonoma married Autopscot, and their descendants became leaders of the community, among them the widely respected

Reverend Benjamin Tashama, who was among the last generation of natives to survive the diseases that wiped out the vast majority by the late 1700's.

"When your dad had the *Wonoma* built, was it a sailing boat only?"

"Yep - didn't have a motor. 1902, maybe one or two boats on the island had them. But in three or four years, he saw what they were doing, so he got one. It was a five-horse, one-cylinder. After that he got a ten-horse, and two main cylinders, and that had a lot of kick to it. They used to say - 'Had different horses in those days'."

"Carl Sjolund told me you had a Model A engine in that catboat."

"Yep. Somewhere in the 1940's I took it over and rebuilt her. I had a dozen Model A engines. They had a kit to make 'em into an inboard engine. When somebody's old car gave out, I'd buy it, take the engine out, fix it all up. Couldn't buy parts. I had them quite a few years. Went dragging with them."

"How long did you have the *Wonoma*?"

"Until about 1978 or 1980. I wanted to rebuild her, but by then they rented the marina for so much, you couldn't get a berth there. I didn't want to leave it on a mooring out in the harbor. Boats get adrift in a storm, you know, come down, might break something, damage something. But I had a dragger anyhow, forty-footer. Went fishing for flounder, fluke."

"So year-round, did you make your living fishing?"

"Oh yeah, never did anything else. There was always something to do. We went dory fishing off the east end. Right off 'Sconset. We had some wonderful fishing out there. Handlining. And then in cold weather we went what's called longlining. Couple thousand hooks."

"You haul them in on a winch?"

"Oh, no. Haul 'em in by hand. Two miles long."

"You mean you just take it with your hand like that? Pull it? Two miles?!"

"One man hauls it in and the man behind him, he coils it up in a tub, otherwise it gets tangled. They'd coil it up, and the hook was in front of them. Then when they go to bait it, they'd pull it up, reach in for the hook, bait it, and put it in the tub. Then they throw the hook over, throw a couple coils of line, do it again."

"How do you keep from getting tangled up?"

"Oh, course you do sometimes. You get some slack line somewhere, then you got a big snake. You get it all twisted up, twisted around. You know when you're hauling them in, the lines are flying all over. You run your hand down looking for the hook. Once there was slack in the line and I missed it. It flew up and stuck in my cheek. The blood was running down my chin, and I hollered at my father and brother - 'Look at me! Look at what I got!' They were so busy fishing, wouldn't even look!"

George has gotten a good laugh remembering this story, but he has to stop to cough for a moment. He is generally short of breath and still hoarse from his battle with the pneumonia, and I have to strain to hear his words. But it is worth the effort. His sense of humor often slips by me, and I realize a moment or two later that he has dryly commented on something.

"It was hard work, but it was fun sometimes," he continues. "We had days out there - haul 'em in just as fast as you could haul 'em. You know, people pay good money to go out on boats, catch fish like that."

"What kind of fish?"

"Codfish."

"Codfish? How big were the codfish?"

"Well they were good size - they were five or six pounds, seven pounds. The biggest one I caught was forty-five pounds. Only one boat goes out from here now. He goes out to the edge of the continental shelf. He catches mostly monkfish. Boats go down there, draggers, got all this equipment on them you know, and they don't see one fish. They wiped 'em out, that's all."

George tells me that the *Wonoma* finally just got too old and would have cost too much to rebuild her. He broke her up and took her to the dump.

"But to go scalloping, I always had a scallop boat," he says. "In fact, to go to Tuckernuck, to fish the shoals, you really need an outboard. I went to Tuckernuck depending on the year - if they were good out there. But most of the time I'd go in town, or Madaket. I'd go where they were."

"Back in those days, were the scallops good in both places?"

"They could be. Yep. There were a few years when they disappeared everywhere. In the '30's. There were hardly any scallops. Just disappeared. The scientists were around here - they decided it was some fungus or something."

"So you had a few years where - did you get any at all?"

"No. We always went out the first day. The least we got was probably about four bushels between the two of us. In the wartime scallops weren't so good. They put a ceiling price on that was pretty low - so I never went much then. But just after the war, the boys coming back - almost all of them went scalloping. Out in Tuckernuck, the grass was just coming back. The dredges would come up almost all scallops. Few little blades of grass, but no trash, no seed. Nothing but scallops. So from then on it was a big thing. Almost twenty years, starting right after the war. They thought it would last a long time. But we'd seen bad years."

"Did you know most of the people in the fleet then?"

"Well - most of them grew up here. A lot of them went to war, but I knew 'em when they came back. Very few new people came 'round here in those days."

"Who did you sell to back in those days?"

"The Island Fish Market - Glidden. When I was a little kid, back in the '20's, my brother and I, we'd go dig a bucket of littlenecks, take 'em over there, get a dollar."

"When you had your dragger, what you were going for?"

"We went after all kinds of things. We went dragging for fluke and flounder. And we went handlining for codfish. We went offshore for them. Codfish and haddock. We went dory fishing off 'Sconset."

"Did you go offshore overnight?"

"We used to go to what they called the Farewell Buoy, fifteen miles east of Great Point. You leave that you go to George's. George's is ninety miles - we never went there. We went to that buoy - and then we'd go from there - usually to the south - that's outside of the shoal. Oh, we'd be twenty-five miles southeast of Sankaty - back and forth. But I knew all the little holes along the shore. We'd go dragging off Cisco. We did that for fluke more than anything else. They were higher price. We shipped them mostly to Sam Cahoon out of Woods Hole. He bought all the fish we could ship to him. They were cheap but he was honest."

"But those long trips were still not overnight trips?"

"No - we'd start out sometimes at four o'clock in the morning - sometimes get back at midnight. You see, with a dragger, where we went, we'd pretty much have the nets get torn up. So usually we stay home the next morning, fix the nets up, clean it out you know, get ready to go again."

"You do that in your fish house?"

"No we'd do that alongside the boat at the dock. We'd mend nets inside. They'd come in a square piece, you know, then you get and put a taper on it. You lace 'em up with certain knots. Lace 'em up down the sides."

"Did you have a crew of guys that would work with you?"

"Nope. Just my brother and I went. Carl Henry once went. Charlie Sayle one summer."

"Who was the first guy you said?"

"Carl Henry. Carl Sjolund - I always call him Carl Henry. He was only a little kid. He started opening in my shanty on the Old South Wharf. Number nine, where they sell baskets now. The buyer was right next door. And it was hard to get openers. So when we were down there, if you couldn't find openers, the dealer'd find somebody for us. Did you know Brownie? When we couldn't find anybody else, we'd get him to open. He was a good opener. But he used to get drunk. Had to have his booze."

When I knew Brownie, he was nearing eighty, with a neatly trimmed white beard, always wearing a plaid jacket and a sailor's cap. He was often photographed because he was very nautical looking. He had died a number of years earlier.

"Did Beinecke own the wharf yet?"

"That's when I left. 'Cause he bought it. Used to be Island Service Company. They had a manager - he was great. He was honest, but oh he watched the pennies. We had an old scallop shanty - we had an agreement with him that at the end of the season we had to give it back in good shape. They always let you tie your boat up. You didn't have to pay for that. You paid for the shanty. Of course the coal for the house, whatever supplies we needed - they expected us to buy it from them."

"Were a lot of those places fishing related then?"

"Oh yeah, the whole place. On the south side the bigger boats, 'cause they had more depth. Beyond the elevator there, you know the big high one, they had that full of nets. In the wintertime go in there, mend the nets."

"No A&P down there?"

"No, no. That was a lumber yard down there. And an ice plant was further down."

"So that was all connected?"

"Yep. The whole wharf."

One of the monumental non-events in Nantucket's modern history took place in the early 1950's. The town was offered Straight Wharf, which had grown dilapidated, for a dollar, and turned it down. It was later bought by a summer resident and heir to the S&H Green Stamp fortune named Walter Beinecke, who also bought Old South wharf, the home of Island Service Company, from the Deeley family in the mid-1960's. Operating under the name of Sherburne Associates, after the original town named for the Gardner's ancestral home in England, Beinecke redeveloped the wharves into what is now the boat basin, bought key hotels and downtown commercial space, and envisioned the elite tourist trade as Nantucket's future. By the mid-1980's, his holdings had increased to the point where one of the early issues of *Boston Magazine* featured him on the cover with the headline - 'No Man is an Island - but Nantucket's Walter Beinecke Comes Close'. Opinions regarding this turn of events tend toward polar extremes - many view him as a benevolent dictator whose manner of developing the island actually preserved its beauty and saved it from a worse fate. Others view him as the person who started the whole tourist industry and wrecked the island thereby.

I asked George if people fished all through the winter.

"Oh yeah. You made a lot of money in the winter. 'Course if they had a freeze-up, sometimes for maybe a month you couldn't get out. The bigger boats that could get to New Bedford could fish from there. But it was pretty hard in the wintertime. Scalloping, we went around the ice sometimes. My brother got in trouble, he got caught up at Coatue once, couldn't get back. I wasn't there but they landed on the beach. I think they finally dragged the boat up on its side."

George stays young by planning for the future. He wants to re-do the pilings under the old fish house, but says he will probably have to conform to new building codes. Anyone who has worked in the building industry knows where this road leads.

"I'd like to fix it up so it would hold together. 'Course if somebody's going to buy it, they might tear the whole thing down."

"Have you had anybody offer to - "

"Oh, yeah! All the time!" George laughs.

George has turned down each and every offer. As long as he is a fisherman, he needs a fish house. And a fish house needs to be on the water. But as time creeps up on him and his century-and-a-half old fish house, George has three choices: spend the time, money, and energy to keep it functional and still charming; let it rot on its pilings and sink into the ocean; or sell it before the inheritance tax forces a sale anyway. Its HBU (Highest and Best Use in appraiser parlance) will be calculated. But since intangibles such as visual charm and historical worthiness cannot be measured by square foot, direct market, or comparable sales comparisons, they will be treated as valueless - as opposed to the appraiser's opinion which will be treated as gospel. HBU will no doubt be achieved when the fish house is gutted, its crooked lines straightened out, its ungainly windows replaced with Andersen thermopanes, and new French doors open out onto a deck instead of a dock. And in another hundred years or so it may be a charming old house again.

December 20

Sunrise: 7:02 Sunset: 4:14
Temps: Low 32° High 52°
Conditions: Clear

gonad index

Dave Fronzuto and Tracy Curley are both in the Marine Department office at the same time, another sign that the season has slowed down. Tracy has just raised sixteen wire cages that have been in the water since early summer. The cages hold mesh bags which were filled with scallop spat from the Marine Department's propagation facility between the Coast Guard station and Brant Point Light, the beacon of arrival or departure from Nantucket. The cages are an experiment to measure how well scallops grow in various locations around the harbor. These are the so-called grow-out zones.

I had just read an article by Tora Johnson in *Time of the Islands* magazine about the Martha's Vineyard Shellfish Group, headed by a shellfish biologist named Rick Karney. The Shellfish Group has been in existence since 1976, and was originally funded by that island's six towns to propagate clams, oysters, and scallops. In 1979 they built a hatchery, which was thought to be the first publicly owned hatchery in in the nation. In 1986, despite these efforts, the scallops disappeared everywhere except the traditionally good grounds off Cape Poge, and had not returned to most areas even ten years later. No cause was found. Today, the Shellfish Group annually provides over five million scallop seed to the Vineyard towns, as well as seven million quahog and four million oyster seed. They have also received half a million dollars in federal grant money to re-train former George's Bank fishermen in aquaculture techniques.

When Dave is asked how Nantucket's propagation program compares, he says that the economics will never make sense on Nantucket. The Vineyard has the advantage of six towns chipping in money.

"For quahogs it makes perfectly good sense: grow them in a hatchery until they're field plantable. We also had a program with soft shell clams where we plant spat and put predator netting over the flats. The netting holds the spat, it reseeds that area, keeps the predators out, keeps the birds out, keeps the crabs out for the most part. You're basically reseeding a productive area. So that works fine for clams. But we are all in agreement that with scallops we cannot grow enough to make a difference." Dave begins to enumerate the reasons the Vineyard's approach cannot be compared to Nantucket, ticking them off on his fingers: "I have more fishermen than they do. I have more area than they do." Then he gets sidetracked as he elaborates: "For instance - 350,000 scallops flown in here the other day from Maine on a charter commercial flight, all 35-40 millimeters big. Would not fit in our boat. And you can't dense pack them, they have to be open. How are you going to grow those out to a size bigger than that? You're not. It's so labor intensive. It's food and water quality critical. You're not going to make a difference on the natural set. We tried that - we tried that for several years. A million scallops, two million scallops, all that big (he indicates a very tiny scallop with his fingers), it doesn't matter. Does that mean that we shouldn't do it? No. But what we've learned and how we're doing it now - in nursery areas, in areas we know that the scallops will grow, and using them to augment, and to find out what's going on with water quality, and growth rates, and all that stuff. We've learned now, through Tracy and her efforts with growing out, that you hold them as late as you can, September, October - the later the better, meaning you've allowed them to get as big as they possibly can. Now, in the sixteen grow-out cages, we had great grow-out. We had very little mortality. Sixteen cages. Sixteen times eight number of bags. Plus the bags that we had on the bottom. It was manageable. We did the grow-out analysis, we dumped those bags - everything's fine. We had scallops this year that were nice 30 millimeter scallops, beautiful scallops. However, we only had about 30,000 of them. But you can't do that with two million, because of the size. If you're dealing with too many, you don't get them up to a sufficient size. That's what we did for years and years and years. The problem is, as we just recently learned within the last three or four months, is that they do not have the protein energy biomass - not biomass, but chemical mass..."

"Gonad index," interjects Tracy.

"Thank you. Gonad index ... to survive the winter. So you dump two million animals that you've paid an awful lot of money for, and you don't know whether they live to the next twenty minutes because you can't track them. And you've just learned that they probably are not of a sufficient size to make it through the winter. Now we know it's better to do smaller numbers, grow them to a bigger size because they are manageable, put those in a nursery area, and then come back and dive on those and find out what's going on with the mortality etc. etc."

"If you dump them just before the season starts - there's no chance they'll get caught?"

"Right - there's no doubt that they're a seed. They're flat, they're shiny, there's no growth on them, no growth ring, it's just a flat, tiny little shell. Even though they're as big as a Madaket scallop that we're harvesting this year."

A few weeks ago I had chuckled to myself when Tracy had mentioned that there had been some talk of attaching tiny battery powered radio transmitters to scallops to track their movements, but the price tag had been astronomical. But from the perspective of saving this scallop industry, and perhaps recreating scallop industries in areas that once supported them, I see now where the knowledge might prove valuable. Dave continues.

"I've had some of the best aquaculture people in the northeast look at our facility over there, and they've all said the exact same thing: 'Dave, it's an absolutely wonderful facility. It is tremendous. It will never be big enough to make a difference. What we've had to do is to realize our own limitations. But we have also had to realize the animal's limitations, and adapt to that. It's fisheries management for more adaptive reasons. Rather than saying 'Close the fishery, open the fishery, do this, do that, run over here, run over there'. We did that. We did that for ten years. And we honestly believed, myself included, that we were making a difference. Until the science proved us wrong. And the Vineyard has come to some of those same conclusions, just like we have. The other side of that story is that I know communities that literally pay their fishermen to go out and harvest as many adults as they can. In one community's case: a hundred and fifty adults. That was it! I mean, these people had tears in their eyes, they were damn proud of themselves. They put them in cages and they protected them so they could get them to spat so they could have some scallops to analyze in the next year. My guys are getting that many in one dredge! So you can see the orders of magnitude that we're apart. We're managing a different level of fishery right now."

Tracy interjects here. "We're probably one of the only communities working on water quality as vigilantly as we are. We're not only looking at the animal, we're looking at the environment that it's living in, trying to do the environmental analysis to sustain the scallop population and to bring it back. We're not just dumping animals into the harbor and saying 'The animals will repopulate and everything will be fine', when there's no other changes going on in the harbor, and where we know there's all this development pressure."

"By having Tracy on staff," Dave says, "She can say about the road runoff, the storm drains, the fertilizer, the build-out, the mooring field, etc. etc. All the pressures that are on the harbor that get you down to what a lot of shellfish officers know, which is 'My guys have got no scallops.' And that's all they know. But we tie it all together. We're on the Watershed Committee, we work with the SHAB board. And like I've said before, I'm not trying to get back to 40,000 bushels, I'm really not. Because I'm smart enough to realize that those cycles that we've gone through for the last thousand years do not have the pressures on the harbor that we have today. So if you can get back to a sustainable catch of 15,000 bushels - some years it's going to be nine, some years it's going to be fifteen; some years its going to be twelve, some years its going to be nine. And that's the cycle within the environmental conditions we have. It will still be cyclical. You will still have the rainstorm on the wrong day, you'll still have some guy polluting, you'll still have some reason why you don't have a success."

Tracy mentions a couple of surprises they had found in the mesh bags. One was a set of natural scallops that had somehow gotten in. But a few of the bags were completely empty, or only contained a handful of scallops. Also in the same bags were a number of cunner fish, heretofore not known to occupy the harbor in great numbers. The cunner fish had clearly feasted on the baby scallops, shell and all, until the scallops had reached a certain size threshold, too large to be prey.

"Think about this," asks Dave, and adopts the tone of a man bringing a new riddle into a pub. "How did that fish get in there? It's about this big, and the mesh is only this big," he says, demonstrating that the mesh bag has a very tiny weave and the cunners are about the size of a pinky finger. The answer, it turns out, is that they got in as cunner spat also, and grew up right along with the captive scallops.

Dr. Belding also studied scallops in grow-out cages in 1910, examining factors that contribute to the growth of scallops, such as depth, temperature, and current. He found that environment seemed to be the biggest factor, with one curious phenomenon recorded when caged scallops are placed alongside natural scallops in the same conditions. The natural scallops always grew bigger. Belding posited three possible explanations. The first, possible density in the cages, was discounted because the same density existed outside the cage as in. The second, that seaweed or other growth on the cage may have slowed circulation, and therefore food supply, had some plausibility. But the third, far more intriguing, was that their smaller size may have been due to 'the lessened activity of the scallop as compared with the freedom of those without, i.e., lack of exercise.'

Perhaps Marty Morse's school of swimming scallops was really just a fitness class.

December 23

Sunrise: 7:03 Sunset: 4:15
Temps: Low 22° High 44°
Conditions: Snow

Snowfall settling on the 'Sconset footbridge. Time stands still on a snowy day.

first snowfall

Late morning brought the first giant cottonball flakes of the season drifting down, and by noon they began to stick. Snow over water always stirs the soul - even the souls of the two dozen or so hardened scallopers still working. But these fishermen know that this lovely scene also signals something else: winter's first shot across the bow. The next three months are only going to get harder.

The majority of the scallopers who began the season are not complaining. It has been a much better year than the last two. They made good money, and who is going to complain about a white Christmas?

And as if the day were not pretty enough already, nightfall brought a full moon which saw its brilliance reflected everywhere, turning scattered patches of fresh snow into newly discovered jewels, silhouetting sharp and graceful pine branches, ringing still, black ponds with neon, and silvering every gray shingle on the island. An exceptionally close lunar approach coincided with the short and dark night of the winter solstice.

Astronomers said it was one of the brightest full moons of the century. From my vantage point overlooking Miacomet Pond, I, for one, was convinced it was so.

From a fisherman's standpoint, such a lunar event creates unusually high and low tides. Rick Kotalac had gathered up a few scallops walking out on the shoal in Madaket, the same shoal we had been stuck on during our day on the water. Rick is taking a sabbatical, as he puts it. In other words, he is effectively done for the season. He has kept his shop open around three hundred straight days now, with only three Sundays off. He likes to get in forty days of scalloping, no matter what. He managed thirty-eight by my count. Indeed, when he's out scalloping, that doesn't mean he quits running Brant Point Marine, he just does both. But it is now two days before Christmas. It is cold. Fishing is hard. Scallops are thin. It is time to rest.

We talk about how it is now and how it was then. Nantucketers like to work. They are enterprising, they are imaginative. But mostly, they just like to work. They like physical work, and they like hard work. But they also like the rest you earn between seasons. Just a few years ago, one could expect a couple of months in spring of moderately hard work if you were in retail or restaurant work, then about three months of simply insane work, then another couple of moderate months, and then four or five months of rest, meaning perhaps working one or two part-time jobs. But one could also count on having some spare time to collect one's life and enjoy it. It was a little steadier in the trades, with spring being the crazy working time, summer and fall being steady, and winter being slack. Now, with Nantucket 'enjoying' one of the hottest home construction markets in the nation, as they say in the trade papers, or 'suffering it', as they say on the streets, tradesmen have been buried all year-round for the last two or three years, to the point that seventy-five tradesmen are flown in every day, since the locals can no longer keep up. The ripple effect keeps the service economy overheated for a full nine months.

Rick and I have talked about how few people under forty have taken up scalloping. The relatively easy money on shore has kept many of the younger generation scallopers away from the water in the 1990's, and has compressed the time when the year-round community can enjoy its own company. But it does say something about the nature of scalloping when working construction on a cold, windy island in the middle of winter is considered 'easy money'.

leilani

With Christmas approaching I called up Larry Miller, whom Tracy Curley had told me is her favorite scalloper, and whom Eddie Loring had mentioned as someone he liked to see on the water. Larry will be seventy years old in May, and has a brand-new aortic valve for his heart. He hesitates to represent himself as the quintessential scalloper. The two men he most admires in the scallop industry are Carl Sjolund and Oscar Bunting.

"They epitomize the business of scalloping and the business of living," says Larry. "Scalloping is incredibly hard to do, incredibly hard to succeed at from a financial standpoint. I could give you a lot of pointers on what not to do," he says good-naturedly.

By the time Larry started scalloping in 1982-83, he had already achieved financial success. Originally a 'Sconset summer resident, he joined the Navy. When he returned to Nantucket in 1954, he became general manager of Sherburne Oil. In 1970 the operation sold out, and he circled the globe in a sailboat for a dozen years. He learned scalloping working with Gibby Nickerson for four years, until Gibby suffered several strokes. Larry then struck out on his own.

"Of all the things I've ever tried to do, scalloping turned out to be the most challenging," he says. "When scallops are plentiful - they're cheap. When they're scarce - they're expensive. Hard to make a living at it. I don't know of anyone with wealth who came to it from scalloping. I had inherited my wealth, so maybe I didn't go at it as hard as I could." Larry tells me he once got lost in fog with a knowledgeable scalloper, and wound up in the surf between Smith Point and Tuckernuck. They had to jump out and get the boat off of the shoal. The engine took on a lot of sand, but lucky for them it still ran. "That was more dangerous than anything I'd encountered sailing around the world," he says.

Larry's boat was named *Leilani* after his granddaughter. He eventually sold the *Leilani* to Bob Cole. When asked if Bob changed the name, Larry says that he hadn't. "Like any fisherman, he doesn't want to do anything to make his luck any worse, so he didn't change the name."

December 28

Sunrise: 7:05 Sunset: 4:19
Temps: Low 26° High 35°
Conditions: Stiff north wind

red flag

The red flag is something I've been waiting for. Scallopers cannot go out when the temperature is below twenty-eight degrees, because the seed might freeze on the culling board before getting thrown back into the water. A red flag raised on the shiny, white flagpole at town pier, or the twelve foot pressure-treated two-by-four at Jackson Point, indicates that the temperature is below twenty-eight degrees. If it rises above that before ten a.m., the flag is lowered. If not, scalloping is cancelled for the day. The flag went up for the first time yesterday, so no one fished. It has just cracked thirty degrees at mid-morning today, cold and blustery, so the flag has just been lowered.

Next door to town pier, Brant Point Marine is closed. Dave Fronzuto had said earlier that about thirty guys are still fishing. "If they are committed, they will go out today, since they missed yesterday."

David Sylvia is in his boat at town pier, arranging his dredges. David was one of the last scallopers I saw before Christmas. He is a chef as well as a scalloper, and had brought Tracy and Sue Mueller, the Marine Department office manager, a holiday offering, some surprise tasty dish, on Christmas Eve. Now, he is sitting at town pier waiting for the red flag to go down. Even at the dock, the *Black Crow* is bobbing up and down. We have to shout across the water and above the wind to speak. I ask him how many other fishermen will be out today.

"Oh, I don't know. Probably some guys got out earlier, because they check the flag first thing in the morning. I had somebody call and tell me. Try to catch something, just to say I went," says David. "Have you ever seen when they have like a shotgun start, they're waiting for the flag to come down?" Dave asked. "It's kind of funny. Well, it won't be so great now, because half the guys aren't fishing anymore. But before, they'd all be having

their boats running, and waiting, and waiting, and waiting, and they're waiting. It's kind of like a race, kind of like the old race, and as soon as that red flag's down they're all - 'vroom!' - going out to their favorite spot, trying to get there before someone else gets to their sweet spot. It's really fun. I used to always think if I did circles they wouldn't know where I was going. Then they'd say - 'That's David!' It's really fun to watch. I was always a slow boat, so they were always ahead of me. I always just saw their rooster tail."

January 1, 2000

Sunrise: 7:04:19 at 'Sconset Sunset: 4:21
Temps: Low 24° High 45°
Conditions: 24° and clear at sunrise

millennium dawns

At 6:30 a.m. I crested the rise overlooking 'Sconset to a sight I am likely never to see again: an unbroken stream of taillights on Milestone Road, a caravan stretching from the cranberry bogs all the way to the sea, snaking along like that unforgettable scene in *Field of Dreams* where Kevin Costner built it and they came.

They were coming from every part of the island and bringing friends from every part of the country who had made a winter pilgrimage to Nantucket - not, by any stretch of the imagination, a typical New Year's destination. Two-thousand people, probably a quarter of the on-island population, were heading to 'Sconset this morning. The reason for all this attention? Astronomer Vladimir Strelnitski of the Maria Mitchell Association, named after Nantucket's 19th-century heroine and first woman to discover a comet (and she was just a girl at the time), had calculated that Nantucket would receive the first rays of new millennium sun on U.S. soil. This contradicted long-accepted wisdom that Cadillac Mountain or Lubec, Maine would have the honor. Strelnitski posited that latitude and altitude would play as important a role as longitude in seeing the sunrise, based on the curvature of the earth.

In the end, a conference call was made among astronomers at each location, and at 7:04:19 a.m. the answer was affirmed - Nantucket had received the first sun. Cadillac Mountain, at 1,534 feet, followed forty seconds later, and Lubec, at sea level, fifty-three seconds after that. It was truly one of the more memorable moments for all those who took the trip that morning. At Codfish Park, people saw the sun rise in waves - first the dozens of folks on the footbridge and bluff above Codfish Park, then the hundreds of us at the top of the beach, and finally the hundreds more right at water level. For a moment the frigid conditions were forgotten as the bright orange sun brought the dawn of a new millennium to Nantucket.

January 3

Sunrise: 7:07 Sunset: 4:23
Temps: Low 41° High 53°
Conditions: Clear, breezy

spanky and the raw bar

It sounds like a rock group, but it is actually Stephen Kania, who goes by the nickname Spanky, and who serves the best raw shellfish on the island. I had sampled his offerings two days earlier at a New Year's house christening. Perhaps fifty, Spanky has been scalloping Madaket for many years. He fishes all winter long, then switches to the summer raw-bar season. The house that was the object of celebration is located off the Madaket Road. The front door opens into a large foyer overlooking the moors, and a string quartet was playing as I entered. Even more impressive, and welcoming, was that the hostess was personally cooking the hors d'oeuvres and preparing the main course, assisted by her husband and children, and their significant others.

At 6:45 on Monday, the first day of the year 2000 that the scallopers can go back to work, I stand on the bulkhead at Jackson Point, its green pressure-treated lumber already beginning to gray, and watch a scalloper row a little yellow dinghy out to his scallop boat and tie it to the mooring, step aboard, pull his dredges off the bottom, lay them neatly on the culling board, untie his scallop boat, start his engine, and go off to sea. He looks as businesslike as an executive swinging a briefcase into an Acura and pulling out of a suburban driveway, yet as far removed from conventional society as one can possibly be. An old man named Richard Whelden, owner of a pleasant, weathered face, mouth perpetually turned up in a grin, is sitting in a pick-up truck watching the harbor. Having spent nearly forty years on the water himself, he is now content to watch his son Howie, somewhere out where the sea meets the sky, fish. Three other boats intrude on today's sharp horizon, and I imagine one of them is Spanky.

January 4

Sunrise: 7:07 Sunset: 4:24
Temps: Low 46° High 53°
Conditions: Overcast

the nub of it

It is not nearly so pleasant at the first SHAB meeting of the millennium. Three scallopers are present, and Blair Perkins, who sometimes unofficially speaks on behalf of fishermen to town agencies, is upset. A warden has dumped another scalloper's box and warned him not to take seed. The growth ring issue has come home to roost.

Only about two dozen scallopers are on the water regularly now, because there are not many scallops in the water. It has become a very long, very cold working day. No one is getting their limit and, as winter wears on and savings dwindle, the large scallops that have only a nub for a growth ring are getting harder and harder to throw back.

Blair wants to know why other factors besides the growth ring cannot be taken into account. He cites other obvious characteristics that scallopers have used for generations to determine adulthood: size, thickness, roundness, and growth on the exterior of the shell. He says science should be developed to settle the question once and for all, and that ten years of study are needed to answer the question, and in the meanwhile the scallops that look like adults should be taken, because they will only die otherwise. He is not alone - many of the oldest and most respected scallopers believe this as well.

Henry Wasierski answers that the Marine Department only enforces state law, and that SHAB only exists to advise the town. Neither can alter state law, which says that the only characteristic by which a scallop is legally an adult is the clear and visible growth ring. Henry says that the only people that can alter state law are the legislators, and they will only do it under pressure from their constituents. He says Blair ought to get a fishermen's association together, like when Henry himself founded the first one in 1972, and go to the state and lobby for it. There is an appeal to board member Steve Estabrooks, a science teacher and scallop biology expert whose research has been published in *The Journal of Shellfish Research*, the definitive source of shellfish knowledge. He can do a test in his office to determine if a scallop is an adult or not. But this is also an old story. Every scallop cannot be brought into a lab to be tested, and the test requires opening and killing the scallop anyway. I have heard Dave Fronzuto say frequently that the only test that can be performed from a realistic enforcement standpoint on the culling board or on the dock remains: 'If you have to look ... throw it back'.

The nub scallop controversy actually began on October 7, 1997, when SHAB made front page news by voting to recommend that the Selectmen change the definition of a mature scallop from one with a 'visible, well-defined growth ring' to one in which the growth ring, in addition to being well-defined, must also be at least a penny's width away from the hinge. The board was trying to head off enforcement confusion that was beginning to be felt by Dave and his staff, Dave saying "If the question arises from time to time, then it's something to look into to be more definitive." Now, two years hence, he would probably be thrilled if the question only arose from time to time.

SHAB addressed the issue on this particular evening because the two divers, Phil Osley and Pete Kaizer, had raised the issue that many scallopers were taking seed. Discussion followed. Ken Kelley, a SHAB member, former town biologist, and longtime scalloper, objected to the 'penny rule', saying that the most of Madaket's adult scallops had growth rings smaller than a penny. "What you're going to do is say you can't get scallops in Madaket anymore." His take: "The problem with the whole thing is the town isn't enforcing the existing seed enforcement and hasn't for ten years." Steve Estabrooks said that some small scallops with growth rings obviously hadn't spawned. The only way to know this is by opening a scallop. He feels it is better to be safe than sorry. Tracy concurred, and further added her belief that many scallops can and will spawn twice. Dave Fronzuto noted that, at $10.50 a pound, it is tough to throw a scallop back.

Two years later, upon realizing that 1998-99 had been such a disastrous season, very possibly the end of the fishery, and in the wake of ferocious dockside battles between fishermen and wardens, Dave asked Mike Hickey, the representative of the Division of Marine Fisheries, the same Mike Hickey who had visited on opening day this season, for clarification of the state's position regarding exactly the terms Blair Perkins had again brought up. The reply, dated January 5, 1999, read:

Dear Dave,

This letter is to confirm our recent discussion concerning seed and adult Bay Scallops. State statute, Chapter 130, Sec. 70 states that any scallop without "... a well-defined raised annual growth line ... shall be deemed seed."

This is a legal definition and may not always reflect biological reality. However, the law allows no latitude, even to the division.

Consequently, the larger Bay Scallops that I observed on Nantucket and that you also sent me that had no raised annual growth line are legally "seed" even though they have other indications (size, epiphytes, spent gonads) that would indicate that they may be biologically adults.

Sincerely,
J. Michael Hickey
Chief Shellfish Biologist

On March 3 of that year, in time for scallopers who needed to decide by March 31 whether or not to buy a license, Dave sent the following letter to the Director of the Division of Marine Fisheries. It is the same letter that Phil Osley told me was hanging on the wall at Dave Glidden's Island Seafood Store where many scallopers sell their catch.

Mr. Philip G. Coates, Director
Division of Marine Fisheries
Leverett Saltonstall Building
100 Cambridge Street
Boston, MA 02202

March 3, 1999

RE: Bay Scallop Growth Line

Dear Mr. Coates,

I am writing this letter to petition the Division on behalf of the Town of Nantucket for a modification of the strict interpretation of Marine Fish and Fisheries M.G.L. Chapter 130, Section 70. During the past two commercial bay scallop seasons I have been in constant communications with Mr. Mike Hickey of your Division and have always found him to be extremely knowledgeable and helpful. However, he and I have experienced an immense problem during these past two years and it is by his recommendation that I am presenting the situation to you for review.

The bay scallop with a small nub located at the hinge or a shock ring located 1/2 to 3/4 of an inch up the shell represents approximately 1/2 of the bay scallops harvested during the past two seasons. Several biological tests have been conducted by Nantucket's shellfish biologists and the State DMF with the same result. The scallops have a "nub" or shock ring, epiphytes, spent gonads, and are of a sufficient size to appear to be an adult scallop, they do not have "a well-defined raised annual growth line" as per the statute. Mr. Hickey's letter dated January 5, 1999 reiterates these exact findings.

Mr. Hickey has been on Nantucket for opening day of commercial bay scallop season for the last two years and has examined hundreds of scallops, documenting; size, location caught, growth ring placement, nub or shock-ring and condition of gonads. These scallops have been analyzed by many and the indications are they may be biologically adults but are legally seed. It has become almost impossible to enforce the statute by literally examining every scallop that is landed. Severe penalties have been issued to individuals that have more than 5% seed in their catch. These seed have had no nub, shock-ring, or indication of any growth line. But again, without scrutinizing every scallop, every day, this task has become unrealistic.

The Nantucket by-law outlining the sale of commercial bay scallop permits requires all licenses be purchased for the following season prior to March 31st. It is my intention that for the 1999-2000 season beginning in October that my interpretation of an adult scallop will **NOT** include:

Use of a magnifying glass to see the growth line
Scratching the surface of the shell with either a fingernail or knife
Inspection of the gonad
Overall size of the shell
Epiphytic growth
Indication of a nub or shock-line

My interpretation of an adult bay scallop will be limited to:

A bay scallop with a well-defined **raised** annual growth line.

The issue has been one of heated controversy for two distinct reasons. Primarily, the overall economic impact to the fishermen and secondly a more long-reaching outcome that the "nub" scallops are maybe being harvested prior to reaching their full reproductive potential therefore reducing the overall annual yield.

Attached please find documentation of what the department and the Harbor Shellfish Advisory Board has done and is planning to acquire additional information. A copy of the letter to the Nantucket Board of Selectmen outlining my concerns and recommendations on postponing the deadline for issuance of next year's commercial permits awaiting your response.

I sincerely appreciate your review of this matter and am willing to meet with you and or members of your staff to resolve this difficult matter in a timely manner.

Sincerely, D.F. Fronzuto

In this particular SHAB meeting, Dave Fronzuto does not take a hard stand, partly because the issue, from an enforcement standard, has been settled in his mind and by his letter, and partly because he feels he does not need to get into the fray today. But he is also aware that many scallopers are not convinced, and that he himself, while perhaps convinced from an enforcement perspective, may not be so sure regarding scallop biology. It will not be the last of the growth ring controversy.

While the 1910 Belding report remains to this day a widely accepted body of knowledge regarding scallop biology and habits, its very first page foreshadowed Nantucket's current debate:

> "During the winter months growth ceases and when resumed again in the spring, a heavy concentric line has formed by the thickening of the edge of the shell, comparable to the annual ring of a tree. This is the so-called 'growth line', which identifies "seed" or immature scallop. Other concentric lines may indicate unfavorable conditions for a certain length of time checked the shell building process."

In distinguishing a 'growth line' from other concentric markings, it is noted that the thickening of the edge of the scallop during its dormant winter months is what causes the distinct 'raised' growth line, while other factors, such as the shock of a sudden transplanting or change in water chemistry, can create a similar, but far less distinct line, which modern day scientists sometimes refer to as a 'shock line'. The growth line in Belding, measured from the hinge, typically ranges from 2/5" to 2 3/5", but generally between 1 1/5" and 1 3/5". Transplanted scallops, according to Belding, 'always show the change by the formation of a growth line.'

Belding points out other factors that are still verified today. For instance, 10 to 20 percent of scallops do live into a second spawning season (20 to 26 months is the operable life span), but what is not discussed is whether they are capable of spawning twice. Veteran scallopers will adamantly hold that a scallop that old is a 'slunker' that will probably die before the season is over anyway, and even if it lives is not as attractive a product.

Pages 36 and 37 of the Belding report yield even more interesting data, again relevant to the day, and perhaps still relevant, but mostly reflective of the extent and limit of knowledge under which the Massachusetts scallop fishery operated for the first hundred and twenty years. Among the items noted is the following:

> "The Commonwealth of Massachusetts was the first to put into force a law based on the growth line as distinguishing the adult from the immature scallop. While we cannot say that every scallop with a growth line has spawned, we can definitely state that every scallop without a growth line has not spawned and is a seed scallop."

Belding's take on the growth ring issue has not essentially changed over time from a biological perspective. And both sides of the debate can prove their cases by it. Belding says (his italics):

> "All scallops *under one* year old *must be protected,* as these furnish practically all the spawn for the following year. Only scallops under this age *need* protection. If the scallop under this age is amply protected, the law has done all in its power to insure the future of this profitable industry. It does no harm to capture scallops over one year old; in fact it would result in economic loss if they were not taken, as nearly all die from natural causes before a second season."

This is Dave Fronzuto's view: "Belding was a brilliant man, but he didn't have to contend with the other stresses. People hand-hauled, there were no power boats or heavy dredges, less pollution, no boat washing, etc., etc. It is a different environment."

"It is a different environment."

January 6

Sunrise: 7:06 Sunset: 4:26
Temps: Low 21° High 42°

free beaches, piping plovers

The Inquirer and Mirror reported that Tessa Mae Whelden, born to Norris and Suzanne, was the first baby of the millennium. Also making headlines, a fierce flu bug has hospitalized some islanders, and has forced a good many to stay home. My mother, visiting from California, required a trip to the emergency room. Flu, short for influenza, is often thought of as a mere annoyance. This can be a fatal mistake. An international Spanish flu pandemic in 1918 killed millions worldwide, including almost half a million Americans. People quarantined themselves in their houses during the height of it; some escaped to the country to wait it out. Even today, close to a hundred thousand Americans a year succumb to flu or its main complication, pneumonia.

There was an inaugural meeting of the Beach Access Management Committee yesterday, designed to advise on laws, procedures, and suggestions relating to beach access issues, their main issue being that of beach driving. Remarkably, while Nantucket is mostly privately held beach (the town only owns about a mile), by custom there has been nearly universal access, with very few private homeowners trying to keep people off. This arrangement has provided all the advantages of a communal style of cooperation, without the messy totalitarian enforcement of Big Brother. But this committee was formed because people realized that an explosive growth in the number of beach vehicles, combined with the annual closing of Smith's Point to vehicles to protect nesting piping plovers, has put new stress on other beaches. With that has come incidents of private owners trying to close their beaches. So far, for safety reasons, the fire department has removed barricades, and the town has fought to maintain its paper road and right-of-ways often forgotten and frequently usurped by abutters as their own. In fact, this is not a new debate. Thirty years ago, in 1968, the same issue arose when hunters and fishermen were suddenly blocked from roads they had traveled for over a hundred years by plantings, wooden barriers, and piles of dirt. A county commissioner found a law that said that any road used by the public before 1854 is public forever. Other roads at issue were rights-of-way along Hulbert Avenue, to Capaum Pond, and beyond the Wauwinet Hotel.

Maine and Massachusetts are the only New England states that allow landowners to own property down to the low water mark. But offsetting this, colonial law predating statehood grants public access to land between mean high and low tide for the purpose of taking fish or fowl. In other words, it is technically impossible for private owners to keep vehicles from driving in that area, or people from walking in it, although one could theoretically keep them from laying on the sand. Luckily it has not yet come to that, as it has in so many other seaside communities.

A popular bumper sticker around the island informs the tailgater that *Piping Plovers Taste Like Chicken*. It is a tongue-in-cheek reference to the otherwise unnoteworthy shorebird responsible for the annual beach closings, not just on Nantucket, but on many other Atlantic beaches. Counter-viewpoints are often affixed to high-mileage used cars, colorful vans, or classic Volvos. My favorite is the green-lettered *Nature Bats Last*. But a scalloper's truck in Madaket proclaiming *Piss On Plovers - Put People First* may signal that the plover debate has escalated beyond a tone of chiding civil disagreement into a new and disturbing level of animosity.

Piping plovers occupy a niche in the natural, cultural, political, and legal world that makes them the point guards in the conservation war. Like snail darters and spotted owls, they are on the endangered species list, and thus come under the purview of state or national regulation. In addition to the natural threats of predation by gulls and other dangers, a favored playground for hatchlings is in the tire tracks of four-wheel drive vehicles, which means many of them getting run over. Following a protracted battle which saw the Board of Selectmen sue the Conservation Commission for jurisdiction, a battle which the ConCom won, the beaches are now closed to driving at Smith Point and Great Point for several weeks each late spring and early summer, while several dozen piping plovers fledge under the stewardship of a team of paid and volunteer observers.

But piping plovers are not the sole object of protection. Like the darters and spotted owls - they provide conservation groups a legal avenue to protect the entire ecosystem they live in. Endangered species become the only mouthpiece through which interested parties can speak. In Nantucket's case,

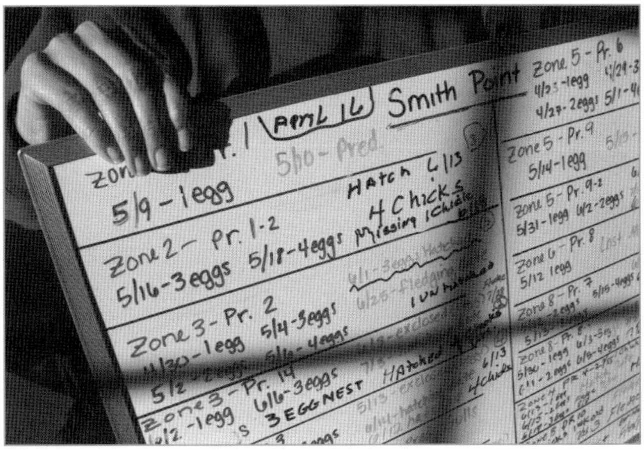

Tracy Curley holding last year's chart of observed numbers and locations of fledgling piping plovers.

these include parties trying to save dune grass and its ecosystem, those trying to save water-dependent shore life, and those merely trying to be at peace with the natural environment.

An unscientific poll would probably reveal that many scallopers are not piping plover fans. On this particular point, many summer sportfishermen, commercial fisherman, and native islanders might agree, favoring the principle of relatively unlimited beach driving. It is also well-known that the latter two groups, regardless of cause or political bent, tend to unite in opposition to any outside regulation of the island.

Plover supporters probably include those summer visitors seeking only a relaxed, unmechanized beach vacation, beachfront property owners, environmentalists, and a substantial population of adopted islanders who frequently take up preservationist causes with energy and passion. In addition to saving the plover, these groups might also favor saving the scallop, if only as a general principle. But in light of recent growing evidence that boat washing soaps cause floating oil from road-runoff and boat engines to sink to the bottom and affect the shellfish, one can only hope we don't see a rash of bumper stickers saying *Nantucket Scallops Taste Like Oil.*

Otherwise unnoteworthy shorebird

January 7

Sunrise: 7:06 Sunset: 4:27
Temps: Low 35° High 47°

comprehensive plan

The preliminary Nantucket Comprehensive Plan appeared in the mailbox of every resident on the island today. It is a 164 page document, produced by six work groups, two subcommittees, two committees, with staff and production teams totalling over 250 contributors, all overseen by the Nantucket Planning and Economic Development Commission. It has been in process since 1995, and contains 106 photographs, 11 charts and graphs, 8 maps, 99 objectives, and a number of sub-objectives. Its chapters include: *Guiding Growth; Housing Needs; Making a Living; Protecting the Environment; Managing Transportation; Island Infrastructure;* and *A Healthy Community.* It is a sweeping, all-inclusive master plan on a scale virtually unheard of for a community this size: an attempt to address the reality that Nantucket has all the problems associated with an extraordinarily wealthy bedroom community of 50,000 or so residents, with the resources of an extraordinarily isolated rural community of 7,000.

So it was with surprise and no uncertain delight that I opened this important and expansive project, which is destined to create months, if not years, of debate, to find a full-spread photograph of family scallopers circa 1955, with the following introduction:

> *In 1980 Nantucket fishermen landed more than 100,000 bushels of scallops, driving an industry that pumped more than $3.7 million dollars into the island's economy. Immediately afterwards, the scallop fishery entered a period of decline from which it has not recovered: harvest spiraled steadily downward - with one brief spike back up - and by 1998 the 100,000 bushel harvest had become a meager 5,000 or less, the $3.7 million returns now barely $250,000.*
>
> *On this, everyone is agreed. On what happened, and how to fix it, they are not. Some say the fishery is simply in an extended, but natural, down-cycle. Others blame pollution: runoff from roads and fertilized lawns, leached toxins from residential septic tanks, detergents and spilled oil from pleasure boats in the harbor. Regulations have been proposed and opposed and lines have been*

drawn in the debate, but as we go in circles one truth remains: there are fewer scallops now than ever before, and the island is the worse for it. With the loss of scallops comes also a loss to our environment, the decline of a traditional Nantucket industry, and the disappearance of money from islanders' pockets.

While this may seem like an odd way to introduce Nantucket's Comprehensive Plan, in many ways the efforts to preserve our shellfish are similar to the larger effort of preserving our island. Scalloping, like many of the challenges that face our community today, represents a cherished part of Nantucket that is diminishing at a frightening pace, with no clear consensus on how to save it. In truth, there is no one action or solution that will bring scallops back: it will be a long process that will likely include hard decisions about waterfront development, landscaping, and the number of boats our harbor can safely sustain. Similarly, there is no magic philosophy that will guide Nantucket's future: we will not find our way by adhering to an economic or an environmental agenda alone, nor by glaring at each other across an ideological divide between the two. Rather, it will take a careful balancing act, understanding how different actions impact different areas, and always drawing the line before one segment tips too far and upsets the scales. Maintaining this balance in the harbor will allow the scallop fishery to be both sustainable and profitable; maintaining it for Nantucket will allow the island to thrive and its people to prosper well into the twenty-first century.

In any event, Tracy Curley had felt the repercussions of Tuesday's SHAB meeting by the following morning. Many scallopers feel that she is the one leading the charge regarding growth ring enforcement, so when the statute gets enforced, it inevitably comes back on her head. In the meanwhile, all have been awaiting new studies, which have just come out this week, and which may alter the accepted wisdom. On Thursday, she and Dave met in Hyannis with the coordinator of the state Coastal Zone Management team, and with aquaculture and shellfish authority Dale Leavitt from Woods Hole Oceanographic Institute, one of the world's foremost marine science centers. Their latest round of tests showed that the small growth ring scallops will spat again, and more importantly, that the second spat will be far more effective, due to the increased gonad size of the older scallop. Dave is now thoroughly convinced that enough of the 'nub' scallops to make a difference will live to spat another season, and that the growth ring will remain the determining feature. The state statute will be strictly interpreted, since there now seems to be convincing evidence to do so.

January 10

Sunrise: 7:06 Sunset:: 4:30
Temps: Low 37° High 55°
Conditions: Overcast, calm

The view from Sayle's Seafood as winter settles in.

sayle's seafood

Sayle's Seafood is as authentic a seafood market as you will find. It sits on family land down Washington Street Extension where the mudflats, marsh reeds, and creeks meet the industrial atmosphere of Grey Lady Marine. Boat hulls lie helter-skelter in the first chill of winter, and the requisite collection of fishing gear decorates owner Charlie Sayle's brother Billy's nearby house. An elegant catboat, perhaps a sister or cousin of George Andrews' *Wonoma*, graces the front yard.

You have to go out of your way to find this place, and wonder if you are in the right place when you do. But if you do manage to find it, you can come in and buy your fish, or get a hot lobster dinner, even in the dead of winter.

On a dreary day just after Christmas I had come to visit Charlie at his fish store, knocking at the back since they were closed. A woman who answered the door told me that he was attending his mother's funeral. When I returned a couple of weeks later, I found Charlie in his office, wiring together slats on the ratline of the *Argonaut,* a seventy year old sloop that he is restoring. A ratline is the web of rope, cable, and wood which holds the mast up and acts as a ladder for sailors to climb into the rigging. This is not the kind of work you can hire someone out of the yellow pages to do. You have to possess the special knowledge these things require, and then sit there and do it yourself. Charlie is doing exactly that. As he does so he is surrounded by receipts, phone messages, ledgers, bills, and all the other paraphernalia of the fish market trade.

To reach Charlie's office, one must walk around the back of the

Charlie Sayle mending the ratline on his seventy year old sloop, Argonaut.

building and past several stations of the long shucking board that he rents out. The anonymous scalloper I met on the first day is still there, this late in the season, opening his scallops. Most people have given it up. He still doesn't much want to talk or have his picture taken.

In Charlie's office, several pieces of frozen flounder, which I mistake for salt cod, sit in a pail on his desk. The wallpaper, many years old, sports a variety of sailing vessels against a blue background. His afternoon shift worker strolls in close to four o'clock and Charlie says 'Try and get in by three-thirty'. 'Just couldn't get back today', says the young man in a strong Irish brogue. Reproach and apology understood, work resumes. I ask Charlie how many scallops he bought this year.

"Nine thousand pounds."

"And it's slowed down now?"

"It's done," he says. "Hundred pounds a day if I'm lucky."

Dave Fronzuto has told me that it looks like the season will shape up to be around twelve to fourteen thousand bushels. This is a significant and heartening improvement, but by a lay person's math, still just above ten percent of the best years, and still at a dangerously low level. I ask Charlie his perspective on the recent down years.

"Ah, it's probably a cycle," he says, echoing most fishermen, and borne out by years of evidence that fisheries are cyclical. "Probably got a choice between the tourists and the scallops," he continues. "There's a lot more money in tourists."

Charlie's family has a seagoing tradition of about eight decades, by his guess. His father, Charlie Sayle, Sr., wrote a column in *The Inquirer and Mirror* for many years called *Waterfront News*. It was a fountain of information that much of the civilized world might, to its lasting detriment, consider trivial, but to the many Nantucketers who relied on the sea for their existence, it provided fascinating, entertaining, and vital knowledge. For instance, in 1968 a complaint was made to the Selectmen that

Shellfish Warden Allen Holdgate should not be allowed to scallop. At the meeting, he replied that he could watch the scallopers better if he was on the water with them. Someone tongue-in-cheeked that if they knew Allen was fishing Madaket, they would go fish town, since he couldn't be in both places. The complainant however, not himself a scalloper, ticked off a list of other infractions overlooked by Holdgate, and demanded answers. To which a Selectman asked if he would be willing to do Mr. Holdgate's job at Mr. Holdgate's salary. The man said he was not interested in the position.

That same year it was noted that fishermen in December were catching eight bags to a thirty foot tow in Madaket, and three bags a tow in town, and that their biggest problem was too much eel grass. Holdgate's main concern was in enforcing his unofficial policy - just don't take any more scallops than you can open. Fishing remained so good that the season was extended until mid-April, and people were still easily getting their limits. No one knows how many scallops were still in the water when the season finally ended. A small item also noted that sea scallops were beginning to be caught in Alaska.

By Nantucket standards, eighty years of seafaring makes the the Sayle family mere beginners, but Charlie, Jr. has made up for at least a couple of generations. He scallops mostly in November now, then has to get back to the business of running his seafood store. Charlie doesn't waste a lot of energy on words. The few he uses, he spits out quickly. I ask if he used to scallop all the way through the season.

"Years ago."

"Who was going back then?"

"Lot of boats. Can't name them. There were like two-hundred boats at one time. At the peak."

"What was the peak?"

"Probably in the early eighties, middle eighties."

"So what happens after these guys shuck, and you get your hands on them? How does that procedure work?"

"We sell to different accounts."

"But first of all, do you pack them in ice and -"

"They're packed in buckets. Then they go into a styrofoam box. Then shipped out."

"On the boat?"

"Plane."

"And how many people on the island buy scallops?"

"Oh - think there's three buyers, Glidden's and Souza."

"So you think you get a third of the scallops?"

"Get my share." He says this with a sly smile.

Charlie originally sold his scallops to Foley. He and Mike Foley have become friends, and Mike comes down to fish with him for the first few days of the season. But Charlie no longer sells scallops to him.

"Yeah, we go direct to the restaurant. All across the country."

"If I were to go to New York ..."

"Go into one of the dealers there. Lockwood is one of them. He's at Fulton Fish Market. Then he sells to the restaurants."

"Any other cities?"

"Chicago. Portland, Maine."

"Also a dealer up there?"

"Brown Trading. They specialize in fancy fish. They sell to restaurants. Real high class stuff. He's been in a lot of newspaper articles."

"How about in Chicago?"

"Shaw's."

"Also a dealer?"

"No, he's a restaurant. One of Foley's best customers. Was." A hint of a grin.

"How about L.A.?"

"Spago. Right in Hollywood."

"Well, that goes to show you, they figure they need to get the best scallop. These scallops are the best, right?"

"I don't think there's any better. Never seen them if there are."

"How about West Coast scallops? Is it much of an industry out there?"

"Naw. They get a few of them. I scalloped out on the West Coast, 1968. Alaska. Up around the Yakutak, Gulf of Alaska."

"Are they similar?" asks Rob.

"No, these are sea scallops, the size of a coffee cup, the eye. They're huge."

"So here in New England, this is pretty much the center of it?"

"This is the best area, right here."

"I've read about Chinese bay scallops. Are those farmed, or do they harvest them like we do?"

"All farmed. Millions of pounds. They're half-Nantucket and half-Chinese scallop. They're not bad."

"What would you think about it if we did aquaculture here?"

"You do that, you're going to lose some areas. In fact, we've already lost some areas, but not good areas. Up there where the oyster farm is, in Polpis Harbor. But they're not good areas. No one's really bitching. Don't know if you can grow anything else there. Is your book going to be on bay scallops, or everything in general?"

"No, Nantucket specifically. But if I claim that they're the best in the world?"

"You won't get in trouble."

We pass a little more time talking about old time fishing and scalloping around Nantucket, when suddenly Charlie turns questioner.

"Do you know George Andrews?" I tell Charlie that I had met him three weeks earlier, and am looking forward to talking to him again this weekend. "How is he? Is he still as sharp? I haven't talked to him in a year or so." Charlie looks dismayed at the news that George didn't fish this season, and that he was recovering from pneumonia. "I used to scallop with him years ago. Very smart man. No moss on him. He's got to be eighty - eighty-two. That's a miracle."

You can't hire someone out of the yellow pages to mend a ratline.

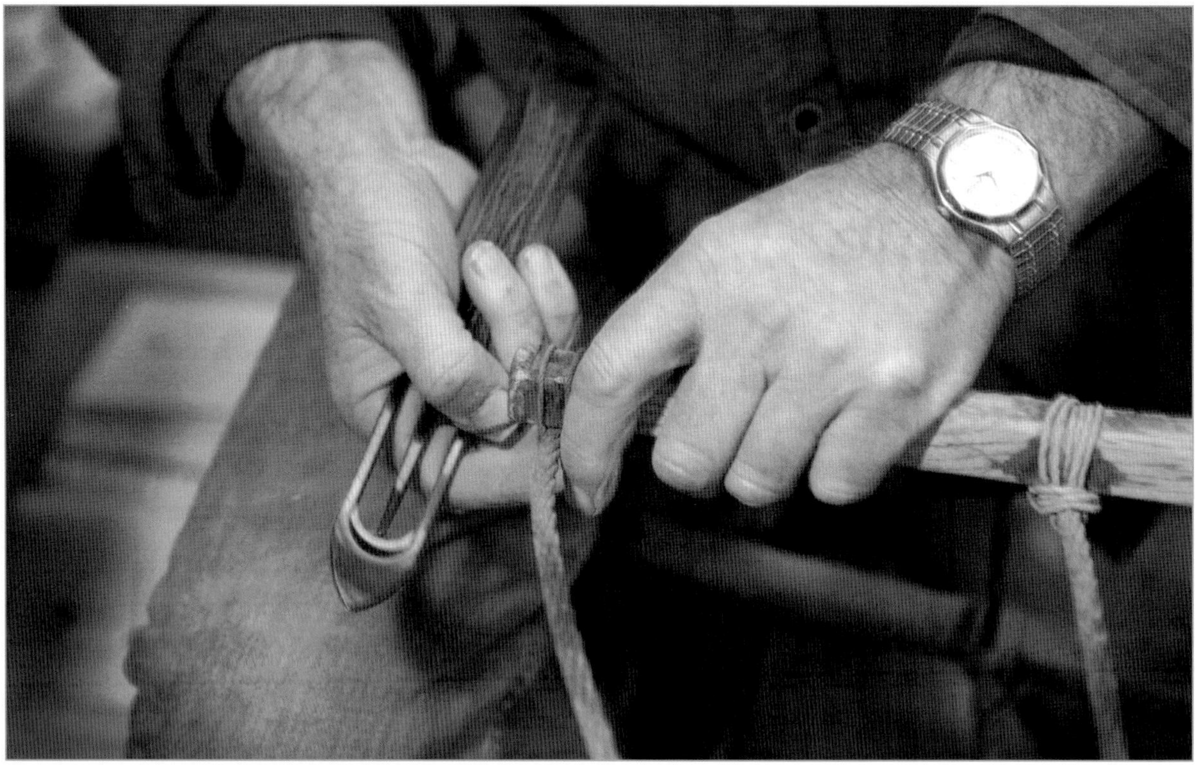

Argonaut

MS 2286 HC

February 2

Sunrise: 6:51 Sunset:4:58
Temps: Low 21° High 31°

wind

Anyone who has spent a winter on Nantucket will verify this: the wind here literally seems to have weight, substance. There is no meteorological term such as 'wind-weight factor' to describe it, nor logic to explain it. But Nantucket wind is, without question, an element both imaginary and palpable, like a character in a play that everybody talks about but who never actually appears. All human endeavor is in some way affected by this wind. Even a completely windless day becomes conversationworthy, for that reason alone. And as if offering themselves as evidence of the palpability of this Nantucket wind, the red flags are getting tattered and frayed. They have flown for over two weeks straight and it will barely break thirty degrees today. Nobody has fished since mid-January.

At one point last week Nantucket had the lowest absolute temperature in the continental United States. On the plus side, cold kills the codium, a non-native seaweed that competes with eel grass, and the red Japanese moss, both of which make the dredges heavier. Some diehards think this makes late season fishing easier, and might go out a few more times when the freeze-up is over. But the fact is there are not many scallops left. For most scallopers, psychologically, economically, and practically, the season is over.

Eddie Loring had told me a little about cold weather fishing.

"When things start freezing up out there, it can get a little hairy. Particularly when they start to thaw out again, the ice starts to move. I wasn't on board at that time, but Malcolm got caught between two ice floes. Fortunately, Smitty came to his rescue and tossed him a line, and was actually able to pull him out and over one of the ice floes. His boat was literally cracking and starting to break up. That's one of the dangers. Because the ice is always moving - it gets in the current. A big piece of floating ice can take your boat with you. I think the divers probably have a harder time with that. Because they may go down - and when they come up they've got a piece of ice floating over them. We've always got to be careful, because even though the divers are out there with their flags set up, they've got a boat with a flag, and usually got a buoy with a flag on it - red with a white stripe - you don't want to get fishing too close around, because you don't know - they could be directly underneath you. He may pop up quite a ways from his boat. That can be a little scary. You don't want to be towing near somebody who's diving. Next thing you know he may be tangled in your dredges. Everybody tries to avoid fishing the areas where a diver might be. I remember the year that a guy took a long walk off a short pier there in town, disappeared. They figured he was floating around out in the harbor. For about two weeks scallopers were dreading it. The worst thing that they could imagine was to find him, to dredge up the body. There were some people making jokes about it, but other people were like - 'This is the last thing that I want to find out here.' Divers are good though. They don't normally go into areas where there's going to be boats, anyways."

"Most of us have beards out there," Eddie continued. "I used to grow a beard every winter just to go scalloping. It gets so cold at times, especially with all the spray, it would literally freeze right to your beard and face. Sometimes it wouldn't get to twenty-eight degrees until almost ten o'clock in the morning, but everybody'd be down there at their boats, six-thirty, seven in the morning, and they'd just sit out there and just wait until the flags come down. All the boats would be together and everybody'd be on the docks, chatting about this, and chatting about that. Just trying to keep warm and wait until that temperature came up. And then the flags go down and everybody'd go out together. And that was one of the great things, part of the camaraderie."

I figure the reason for Eddie's slightly different take than David Sylvia's on waiting for the red flag might be that the *Blow Me* was a faster boat than the *Black Crow*.

groundhog day

Punxsutawney Phil comes out of his Pennsylvania hole today and tells the rest of the country whether or not winter is over. Quinten the Quahog comes out of the water and does the same for Nantucket. Of course, clams cannot see their shadows, although scallops, with their many blue eyes, could. Quinten, instead, either squirts left or right when opened, which is done, with great ceremony, by none other than harbormaster Dave Fronzuto. Quinten squirted left, meaning six more weeks of winter.

This was not the news people wanted to hear. Last week's headlines were again dominated by the deep freeze, which has disrupted the ferries, caused a child who was injured sledding to be flown off-island for treatment, and prompted the Coast Guard to warn that their capabilities were severely hampered by ice pinning their boats to the dock. Snow-covered ice stretched in a solid sheet from Monomoy to Brant Point, and as far toward Wauwinet as the eye could see. Frozen spray transformed the stacked boulders of the Jetties into a forbidding, Himalayan landscape. Slick gray seals somehow managed to balance easily on the slippery jumble, their home until spring's warmer water chases them north.

February 11

Sunrise: 6:41 Sunset: 5:09
Temps: Low 36° High 42°

civic duty

Rob and I were poring over photographs, warm and cozy in his 'Sconset living room, which, like his house, was built entirely with his and his wife Carol's hands, when a knock came at the door. It was Gerry Eldridge, the redoubtable 'Sconset Water Superintendent, holder of shellfish badge Number 1, and object of Rob's envy. Rob and Gerry engaged in genial conversation for several moments in the doorway, just out of earshot, and when Rob returned he had a somewhat quizzical, somewhat worried look on his face.

"Gerry wants me to run for 'Sconset Water Commission," he said. "I have to take out papers by Wednesday. This is the last thing I need."

Gerry had assured Rob, who is already a member of the 'Sconset Volunteer Fire Department and thus answerable to their emergency walkie-talkie twenty four hours a day, that there is not much to the job and that no one else was planning on running for it. Gerry further proclaimed that the 'Sconset community, of which Rob is such a popular and stable part, would benefit greatly from his services, indeed was in dire need of them. Rob returned, having told Gerry that he would think it over, but the concern on his face was palpable. Although he could not identify any particular suspicions, he nonetheless had a bad feeling that this would be like signing up for some kind of lifelong, community-service prison sentence. I suspected Rob would file papers by the deadline, and indeed he did. Gerry was pleased with an excellent job of salesmanship.

Valentine's Day

Sunrise: 6:37 Sunset: 5:14
Temps: Low 37° High 50°
Conditions: Clear, calm

a hundred years too late

The deep freeze has finally begun to thaw and the ice has broken up enough to fish, but today is the first truly nice day. It is perfectly calm, probably in the mid-forties, simply beautiful. A random scalloper or two interrupts the arctic look of the harbor, scattered ice floes drifting like marshmallows.

Rob went out with fellow 'Sconseter Doug Smith, who lives in a fishing shack on a shell lane heated by a wood stove. Fishing gear occupies his lawn, and a shotgun rests against richly darkened pine walls. Smitty was one of the characters of the Gentlemen Tow, and is one of about two dozen scallopers who regularly work the water these days. Spanky still fishes Madaket. Andy Roberts, Blair Perkins, Neil Cocker, Kenny Duce, John Kittila, Herky Stojak, Jim Nickerson, Fred Conner, Marty Mack, Rich Holdgate, Mike McInerney, Rhys Bender, and a few others go out when they can. Dave Fronzuto told me that each fisherman has a pattern, depending on what else they do for a living. It is unusual for them to fish on Fridays this time of year, or if they do, they'll come in early. It is that way with a lot of the outdoor trades in winter.

"I was born a hundred years too late," says Smitty. He is the kind of man who will say what is on his mind without hesitation. "I love to hunt and fish and I see it going away - " is one of the thoughts on his mind today. Smitty has tried to make his living on the water since his first season fishing. "Back when we were getting three-fifty a pound for scallops, I'd always keep a rifle handy," he says. "I could get thirty-five bucks on the black market for a Canada Goose. That's the way of a waterman."

Smitty bought his first scallop boat in 1977, a sixteen foot Lyman, for $75.00 which he saved from unemployment checks. "It was sitting on the bottom, by Children's Beach," he says. Smitty moved to Nantucket from what he describes as a lonely life on an Ohio farm. He was eighteen years old, and rented a room in 'Sconset from Grace King, where he met Eddie Loring, who began scalloping around the same time. "We were two hippie bricklayers making $3.75 an hour." Many people began scalloping at around the same time, for about the same reasons. Scalloping is an excellent way to make a short-term living, despite the hard work.

Smitty's dog, Gus, short for Argus, has a nose for scallops. His predecessor, Moonshine, used to jump from boat to boat with a paper bag carrying beer to thirsty scallopers. He could even buy the beer. Smitty would send him to the 'Sconset liquor store with a five dollar bill in his mouth, and he would return with a six-pack. Smitty didn't say whether they ever checked Moonshine's I.D.

Smitty will be the first to acknowledge that he was a drunk and a rabble rouser as a young man. Like Malcolm and Eddie, he was fiercely territorial about his fishing space. The idea of an 'open access fishery', that the sea is everyone's water to fish, or if need be, to defend, is central to the mindset of a fisherman, and Smitty was no exception. He would scream and curse, threaten and bully anyone who jumped on his tow. And he made his share of enemies.

"One morning I found a bullet hole in my engine."

"And that didn't stop you?"

"F---, no."

But Smitty has mellowed somewhat. At one of his favorite spots, he was surprised to find another scalloper already fishing there. He was even more surprised when the man started cursing and turned the bow of his boat toward him, in ramming position. "Pulled a Smitty on me," he laughed, but took an uncomplimentary view as he continued. "He's been taking seed for five years and getting away with it. Keeps his boat in Polpis Harbor so no one sees him unload. Today I saw him working an area I fished out before the freeze - nothing left but the seed I dumped back. One f---ing idiot, one bad apple can spoil the whole bunch."

Smitty and Gus

smitty's i.r.a.

Smitty got an aquaculture grant in 1997 to grow quahogs, steamers, and oysters. It was the second such grant given on Nantucket. Aquaculture grants began in earnest in Massachusetts on Cape Cod in 1989, sponsored by the state through its SouthEastern Massachusetts Aquaculture Center (SEMAC). It has since blossomed into a four-and-a-half million dollar a year industry, employing three to four hundred people. Amid successes come occasional problems. In 1993 a disease akin to a black plague of clams wreaked havoc on the industry. The density with which aquacultured shellfish are grown, and the lack of genetic diversity, caused unforeseen problems. Throughout New England, as aquaculture gains a foothold in shellfish as well as finfish, worries about introducing disease into the wild stock and encroaching into open access fisheries has created sometimes heated debate.

Smitty obtained four acres on the northwest side of Polpis Harbor and enough money to plant 500,000 shellfish. Getting the grant meant following a red tape trail not many fishermen are willing to put themselves through. First, he worked through Dave Fronzuto to get a contract with the State Division of Fisheries and Wildlife. Then he had to apply to the town and get permission from the Selectmen. Once that was obtained, he had to file paperwork with the state and with the Army Corps of Engineers. Then he had to pay a $281.00 fee to the state and file a notice informing all abutters of his intention. Then the state inspected the property. If they had found more than one clam per square foot, it would have been designated for recreational or commercial use, and he wouldn't have been able to farm it. They found no clams, and no eel grass in the area, either. In other words, Smitty got the aquaculture grant only because he planned to grow shellfish in an area that had already proven it could not support it naturally. As Charlie Sayle had said, nobody's complaining about aquaculture yet, because none of the good areas are being used.

Which doesn't help Smitty, who plans to use this as his individual retirement account. His lease option is for twenty five dollars per year, per acre, for three years, then five, and then eight, as long as he keeps working the area. Smitty has become both politically and scientifically knowledgeable over the years. He is aided by the fact that his brother is Director of the Marine Biology Department at the University of Georgia. Smitty has applied for a SEMAC research and development grant, hoping to use Georgia and Australian aquaculture programs as models. He has taken the time to study the biology of the scallop and the harbor. He believes in cycles, and his theory is that weather is the primary culprit in the demise of the scallop, especially the rainy summer three years ago which probably killed off a lot of spat. But this is followed closely by his belief that oil sinks into the water column through boat washing. He cites the amount of oil and kerosene in the ground near the boat basin from decades of industrial use. "Back in the eighties a worker got second and third degree burns when his torch hit the sand - there was so much oil in it that it exploded. That leaches into the harbor and sinks. Plus in a rainy year you get more runoff from the roads. And the steamship kicks up a lot of that silt, which gets into Pest House Channel and goes toward Monomoy - goes the wrong way. It used to be ass-deep in there with scallops. No more." He rattles off grey water, pesticides, fertilizers, and two-stroke engines as culprits too. "I looked at algae from an algae bloom through an electron microscope - it looks like a giant piece of pollen. But it is so large, the microscopic scallop spat can't actually get it into its mouth. It starves. One tiny little act can go a long way," Smitty summarizes. He says to put a feather in Dave Fronzuto's hat. "We needed him here twenty years ago."

Meanwhile, Smitty has nearly twenty-thousand hard-earned, brick laying dollars sunk into his underwater IRA: half a million clam and oyster seed. "And I haven't had a clam for dinner yet," he says. But if a bullet hole in his engine didn't stop him, neither will a couple of bad years at an aquaculture experiment.

Smitty preparing to haul in, using his donkey engine. On the great old schooners, donkeys were sometimes used to turn the giant capstanshaft which hauled the anchor. Though steam power made the onboard animals obsolete, their legacy lived on. A captain who needed to insult an engineer might have called him a donkey.

taylor bays

Recalling a previous conversation with Dave Fronzuto and Tracy Curley about the role of propagation in helping the fishery, Dave reiterated his earlier conclusion. "Growing two million scallops is not the answer. Growing two million scallops and putting them in the right areas at the right time at the right size is a help."

"Yeah, is a *help*," Tracy emphasizes. Both understand fully that it is far from the panacea many would like.

"But it's not the answer," says Dave. "If that was the case, then we ought to just stop fooling around and just do everything with aquaculture. However, aquaculture is so labor intensive that if you're not out there every day doing it, you're not going to make a go of it." Dave tells me of a brand of aquacultured scallops called Taylor Bays. "A guy named Taylor took seed from the Vineyard and grew them out to 35 millimeters and sold them. People ate the whole thing, with garlic and butter, and that was the marketable product. The way I ran into Taylor was in 1990, at the Boston Seafood Show. Nantucket fishermen got together and we went up there and had a booth. It was great fun. However, I was walking along one of the aisles and there was a thing from Taylor Bays and he had Nantucket Bay Scallops in bowls and people were eating them. And I said 'Oh my God, it's seed! And from Nantucket!' Because they look identical to our scallops. So I did a little research and found out that in an aquaculture product you can sell them whenever you want to sell them. So the legal ramifications don't matter. But is that where we're going?"

"I don't think so," says Tracy.

Dave thinks that, even though quahogs are propagated and a dozen guys are making some money quahogging, they are still probably quahogging on the natural set. But he also says the propagation programs related to quahogs and steamers is a far different story from scallops.

"The trays or sand rafts or other clamming and steamer techniques - okay," he says. "Doing scallops in the lantern nets or the pearl nets is so labor intensive. We discovered if their densities were wrong, you'd have to cull them, have to move them, you've got to put in more bags, more long lines. We've done it and it is an absolute nightmare."

"The other thing that is my concern with aquaculture is genetic diversity," says Tracy. "Dale Leavitt was here telling us that he's doing a project with the Indian Blanket quahogs to find out how much of the aquaculturally grown stuff is out there and is affecting the wild population. They're having big problems with disease affecting the wild species, whatever it may be."

Dave tells me that the state had contracted with Taylor to provide three hundred million scallops, which they were going to use in tracking and monitoring studies. Taylor wound up providing roughly half of that, and they all died. "You know what my answer to that is?" Dave says. "They should have eaten them before they threw them in the water. When you take sixty animals and make two million - " starts Dave - "You're going to have problems," finishes Tracy. She had pointed out in an earlier conversation that genetically similar animals also share the same vulnerabilities to disease, which diversity can mitigate. Some genetic problems are slip-hinge, brittle shells, and small size.

"Let me run this by you," demands Dave, clearly indicating that this is a rhetorical statement. "The reason we don't have bay scallops in the harbor is because we have poor circulation. And we've spent all this money now on circulation and all this money on water quality testing. You like that theory? Fine. You've got Madaket Harbor. Open to the sound, open to the ocean. Probably some of the best circulation going. Why aren't there any scallops? The other theory says the Madaket scallop's a smaller scallop - maybe it has something to do with the genetics of the whole thing - a lot of small scallops making a lot of small scallops. I don't know. There was a lot of big scallops in Madaket too. And in town, they're freaking enormous. Enormous! Four-and-a-half and and five inch scallops." The term for this is compensatory growth, and it is not far removed from Belding's belief that the environment makes the biggest difference in size. In short, where they are thick, they do not get enough food. "But," summarizes Dave, regarding all the blind alleys that the learning process has led down, "You don't say - pfft - we're not going to do anything."

February 17

Sunrise: 6:33 Sunset: 5:17
Temps: Low 18° High 38°

gold medal

In this week's news, islander A.J. Mlezcko, who thrilled us by leading the first ever U.S. Women's Olympic Hockey Team to the championship at the Nagano Olympics, got engaged. Her picture is on the cover of the telephone book, surrounded by kids on the steps of Pacific National Bank, gold medal on a red-white-and-blue strap dangling from her neck. She allowed each and every one of them to touch it.

Also, after a hundred years in the same location on Steamboat Wharf, Glidden's Island Seafood moved out of town. Rob and I already knew this - we had dropped by to photograph the inside of the store only to find it empty. Sawed-off pipes where the sinks had been were the only hint of its history.

Jalopy busted the scalloper Smitty was talking about. His mistake was to come in with five boxes at two o'clock, after showing up every day at four-thirty with two-and-a-half, like everybody else. Jalopy had been Sherlocking him all week and didn't want to divulge how he caught him. Two other guys got warnings, including the anonymous scalloper.

There was a Marine Department notice up at Carl Sjolund's shanty - SHAB is offering a predator bounty - five dollars a bushel for starfish, green crabs, and mudcrabs. There was also once a bounty on seals - five bucks a nose. Conch and horseshoe crabs were not included in the bounty, because conch is a state regulated fishery, and horseshoe crabs are used medicinally and as conch bait.

Seven boats are moored off town pier, quietly pointing into the wind. A number of dinghies and the Petrel Landing platform are half sunk. The harbor is as empty as you will see it. A beautiful old schooner rests at Swain's Wharf, as though the past century never happened.

March 3

Sunrise: 6:12 Sunset: 5:32
Temps: Low 33° High 48°

town warrant

The Inquirer and Mirror printed the Town Warrant articles yesterday, with the following large-print, two-thirds page entreaty by the Board of Selectmen urging Nantucketers: *"To act upon and transact any business relative to the foregoing subjects which may, then and there, come before said meeting. Hereof fail not and make due return of this Warrant with your doings thereon to the Town Clerk at the time and place of meeting aforesaid."* In other words, show up and put in your two cents worth. Nantucketers seldom need encouragement in that department, but such notification is probably required by law.

Any citizen of Nantucket who can round up ten signatures can submit a warrant article, and it can be enacted into law at the annual town meeting in April. These meetings last for several days, and are marked by raucous debate, inside jokes, feuds, a cast of characters that Hollywood would die for, and a frightening concept to the political establishment: direct government by the citizenry. If a person can stand up in front of three or four hundred fellow citizens and persuade them that their article is a good idea, it becomes law. No middlemen, no lobbyists, no bureaucrats, no patronage, no special interests buying off legislators or ramming lucrative pet projects through for favored constituents.

Real town policy comes out of these meetings. Sometimes the articles involve purely practical matters, like traffic flow patterns, or the placement of a stop sign. Phil Osley had told me that me that diving was outlawed on the Vineyard, and that years ago a warrant article had been proposed to outlaw diving here.

"A group of dredgers didn't want divers out there," he said. "They thought the divers would take over their spots - get into an area and no dredgers could get in. So they put an article in the warrant. But we got a lawyer, got up and said there's nothing wrong with diving. The finance committee voted to disapprove it, because they said it discriminates against divers. It was defeated soundly. 'All in favor' was like, fifty. 'All against', like two-hundred."

Sometimes warrant articles are whimsical, like the time scalloper Steve Scannell took one out to declare himself 'Village Idiot'. Or the time Michael McHone took the microphone and proposed to his now-wife Nancy. Only one vote counted that time.

Some articles remind us why government exists in the first place: to formally decide how we, as a society, will treat each other as individuals. Ostensibly a forum to shape public policy, town meeting is really a yearly examination of our collective conscience. For instance, the issue of legalizing domestic partnerships for town employees was debated in the press, on Main Street, and for three consecutive years it was argued ferociously, eloquently, passionately, and sometimes bitterly at town meeting. For three consecutive years it was voted down, but by ever decreasing margins. In the fourth year, it passed into law with hardly a murmur. The collective heart of the community had gradually been changed. Two years ago, several measures to limit cars coming over on the Steamship passed overwhelmingly, as well as a measure to ban any more paving of roads. That same year, twenty-five million additional dollars was voted to accelerate the town's purchase of open land for conservation, which had begun with the founding of its first-in-the-nation Land Bank some years earlier. That money purchased 348.62 acres. In these regards, Nantucket has consistently put its money where its mouth it, and as a result has devised several techniques to limit development, a difficult challenge in a harsh legislative environment. Fortunately for the reigning national dollarocracy, few towns still operate under this kind of true democracy, so the threat of the common citizen exerting undue influence over his or her own government is limited.

This year, there are 103 articles up for consideration. Of these, 62 were submitted by the town, and 41 by private citizens. No less than seven articles relating to the harbor have appeared. The most controversial was written by a scalloper, former SHAB member, and now candidate for Selectman, Steve Bender, proposing a total ban on boat washing soaps. His contention is that surfactant soaps, which include virtually all soaps, biodegradable or not, break up grease and oil that might otherwise float on the surface, causing them to sink to the bottom and affect the bottom life, in particular the microscopic scallops who are just days old and struggling for survival during the part of the summer when the harbor sees its highest boat use.

global warming

Bender's article is on the mind of scalloper and charter boat operator Bobby DeCosta, who has stopped into the harbormaster's office to ask Tracy if an anticipated study on the harbor flushing rate has come out, because he may want to stand up at town meeting and talk. He believes that global warming is affecting the harbor, and that it no longer recycles itself to the ocean as quickly as it used to.

"I keep a log of the temperature in the harbor," Bobby says. "We got to eighty-one and a half degrees this year. I just want to know if there's something that says where this water goes. Does it go way up harbor, or does the majority of it go to Brant Point and out the Jetties? My personal experience has been that the surface water goes out. I just wanted to make sure that if I got up and said all this water is going out of the harbor, I didn't have a study there that says it doesn't."

Bobby knows from observation, by fishing and watching the direction of the boats in the mooring field, that the tidal flow forks into two currents at Brant Point and at First Point. He also contends that the Jetties have gotten lower over the years, undercut by storms. This he has observed by watching an old movie filmed on Nantucket called *Cheaper by the Dozen*, in which the Jetties are clearly much higher. I ask him when he began to see the decline in scallops.

"It all started when they found those scallops outside the Jetties - they only had small growth rings. They said - 'Well, that looks like an adult to me'. Then they started taking them from inside the harbor. I go diving, and it's still a healthy bottom. Everything's there but the scallops. The crabs are there, the other shellfish are there. There was an article in the *New York Times* about a month ago about the Pamlico and Albemarle Sounds off North Carolina. From all the hurricane runoff, they've got about eight inches of new sediment in the estuary, and about seventy percent of it is fertilizer. When it heats up this summer, they figure it's going to be such an algae bloom, it will kill all the crabs and all the baitfish. They predict the next ten years of inshore fishing in that area will be devastated."

Tracy says that the Head of the Harbor and Quaise are showing signs of eutrophication right now. "The phytoplankton composition, species composition is changing over time. That's bad for the scallops."

"Due to global warming and lack of circulation?" I ask.

"Due to too many nutrients going into the harbor."

"From washing boats?"

"That contributes. But it's all through the watershed, the estuaries coming in. Temperature is a factor too. It's warmer for longer periods of time. Which is also bad because you have a higher chance of having an anoxic event, which is re-suspending all the nutrients."

"There's less water circulation," says Bobby, coming back to his original reason for being here. "Global warming I think is a major part of it. It's changed the whole cycle of the scallops, the time of year that they spat. Fifteen miles south of Madaket the water was eighty-six degrees this year. They're catching dolphin and other southern fish. Kids that I know were catching triggerfish and parrotfish. Tuna fish were in the rips at Great Point, Sankaty, and Quidnet, which has never happened. The water was so hot we saw flying fish in the Chord of the Bay this year."

"That's Gulf Stream water coming up?"

"That is global warming and nothing else. The water heated up to eighty-one degrees in the harbor this year. I've never seen it break seventy-five. You can't have water that hot and not have it affect the scallops. These are cold water animals; they're not the Florida version of the scallop. An example of how global warming has affected this island: in the years that I grew up here, in the wintertime you had a ring of slush ice around the whole island. From around the first of December to the end of March. It would go out around a quarter of a mile. It wouldn't

"Flying fish in the chord of the bay."

freeze, but it would keep waves from breaking. When you'd get a big winter storm, the slush ice would act like a wave break. By the time it got to the beach, it was just a gentle swell. You lost a few feet of beach to erosion, but you didn't get these slamming winter storms that you get now. The last cold winter that we had was ten years ago. I was building my house. I remember the only place we could scallop was Madaket Harbor, because town was pretty much frozen up. But the harbor was good and healthy come spring too."

"Another thing is the fact that for five years we took the little growth ring scallop. I'm not a biologist, just a fisherman, but you take a scallop that's an eighth of an inch big, how much spat can it physically throw out? We always left that scallop, we never took it. You know, come February and March, that scallop was just automatically assumed a seed. Because there was enough of the big growth ring scallops that we never had to take it. Off of Folger Point, I can remember years when you could throw a dredge and tow it a hundred yards, and come up clean to the bar with that scallop. Every one of them had a growth ring, but you would just dump them back over, because every one of them was a seed. When those scallops spat next spring there was tons of spat in the harbor. Some fishermen can be their own worst enemies. I mean, last year at the public hearing we fought and fought and fought to not take that little growth ring scallop, and a handful of guys got up and said 'You're putting me out of business.' And they won, you know."

"Some guys tow way too long," Bobby continues. "When I learned to scallop twenty years ago, I learned from old timers who'd been doing it for years. You didn't tow for thirty minutes. You towed for five minutes and then sat there and culled off. You used six pound weights the first two weeks of the season, then the rest of the season you took your weights off. Now you've got guys that don't understand the logistics of how the harbor is. They don't understand that scallops get in rows, that scallops get pushed up against banks. They don't read the contours of the water or the grass patterns. They just throw their dredges, tow in one direction for twenty minutes until their culling board is clean, then haul back and dump the stuff. Then turn around and tow back the other direction. You know, guys used to put out buoys, we called them cull buoys, and everybody would congregate around the buoy. And all the crap that you culled out would be piled in that one spot. So you'd have clean scalloping over here. When the grass was thick, you'd hold your warp and you'd feel it bouncing, you might only tow fifty feet and then you'd feel it slide. You'd stop and haul back, and the dredge would only be a quarter full, but they'd be just scallops and grass. Now these guys just keep right on towing until their culling board's clean. They don't get any more scallops in the dredge, but that dredge just digs in and rips the grass up by the roots. They think they're getting more scallops, in reality they aren't. There's a definite science to scalloping, some of these guys just don't know it. The guys that do, they're in at eight-thirty, nine o'clock every morning. The guys that don't are still out there at four o'clock, drilling away."

"There's also definitely nowhere near the eel grass in the harbor that there was years ago. We have, we call it slime grass, that real light mossy stuff. When you dive and you look at it, it lays right in amongst the eel grass. It chokes off the light to the base of the eel grass, so the grass can't really grow. When we had cold winters and ice, that cold, we called it winter blight, would kill the slime. You'd see it after the ice melted; all your dredges would pick up this black goo."

"That's a macro-algae that does better in high nutrient situations than does eel grass," says Tracy. "Eel grass thrives in low nutrient situations. We found that a lot of the nutrients are coming through Nantucket Sound. It's a regional problem. But we're trying to work on our little section. The problem is even if you stop the fertilizers today, you've still got ten years' worth that are going to keep leaching into the harbor, because that stuff lays in the ground for a long time before it finally filters in through the groundwater. And that's our fear, because we're already seeing the water quality worsen."

"It's not just pollutants that are killing the eel grass," continues Bob, "Because if it were, you'd have lots of eel grass in Madaket Harbor, especially up around Tuckernuck and Muskeget where we used to have acres. It's not from sanding over either, because the bottom is pretty much stabilized there. The shoals have not really moved more than a few feet in a year. Yet you go up behind Tuckernuck, the eel grass used to be so thick it would choke out an outboard. It would grow right to the surface. That grass isn't thick like that anymore."

"Look at how many more boats are in this harbor now compared to five years ago. Every year more and more work up into Hussey Shoal, and they're anchoring at First Bend now. That never happened. You can't close the anchorage, because this is considered a 'Safe Harbor' by the federal government. The Jetties is now the overflow anchorage, there were never boats there. One anchor can do a lot of damage. A proper anchor is going to have at least twenty feet of anchor chain in front of it, you sit there and you swing 360 degrees and that lays on the bottom and drags right over that grass. Then you take a good year like this where the scallops were good in Madaket in the shallow water, you've got twelve-hundred family licenses. I would say that if they harvested conservatively twelve-hundred bushels, which is very conservative, they probably crushed seventy-five hundred stepping on them."

"SHAB has spent a lot of time trying to figure this out," says Tracy. "Limit it to one per household? Limit it to residents? We raised the prices of them ten dollars across the board and we got so much flak from local people that the Selectmen overturned it. But then, they're everybody's scallops," she notes. There is a word for this concept in regulatory lingo - 'Open Access Fishery.' It is one of the remnants of Colonial law and one that even international treaties wrestle with. Icelandic fishermen once attacked the British Navy over the issue. But in an era when growing regulation and control are inevitable, the concept of clinging to a phrase like 'Open Access Fishery' holds appeal.

"You do have people that come here specifically in the fall to go scalloping, they do cheat, they go every day, they get their scallops, freeze them, and take them with them when they leave," Bobby says. "It's hard to police when you've got twelve hundred people doing it and you've only got half a dozen wardens. You can't be everywhere checking every person. Then, there's a balance. We're a tourist community, a water-related tourist community, and all these people that are coming here on their boats spend a lot of money on this island. It filters down to every one. They stay in guest houses, they eat in restaurants, they buy in the shops, they ride in the taxis, they really do everything here. And to say we're going to eliminate this for a hundred and fifty guys who go scalloping? There's a fine line, it's a juggling act. The boat basin polled their customers from last year, a hundred people, if they couldn't wash their boats at all, would they come back. And all but two said no."

"But who knows - there may be a hundred in line that will," says Tracy.

"I doubt it. And then it's a matter of who's going to enforce it? And is it legal? The federal government says that these are environmentally safe products to use, and they approve them in all other areas. These people have got lots of money and lots of lawyers. All it takes is one good class action suit, or somebody to just wash their boat and say 'Take me to court.' You know ninety-nine out of a hundred judges are going to say this law isn't worth the paper it's written on. My experience over the years is that most boat owners are environmentally conscious. They're really good about trying to do everything they can. Look at the amount of garbage that comes into that dumpster every day from the anchorage. People in the anchorage aren't washing their boats to speak of because they don't have that much water. How many people come and pump out voluntarily? Sure there's always the guy that's going to do it. But this is the only place I've ever been that people don't clean fish and throw them in the water. All the charter boats leave our filet stands open so that anybody who cleans fish comes over and puts the carcasses in our barrels so they don't get dumped back in the water. But if you were to say that you can't wash your boat, then you may as well say that all the storm drains should be sealed."

"Which they should," Tracy says, predictably and unhesitatingly.

"Or you can't wash your car. You shouldn't fertilize your lawn, because all that stuff is the same event, ends back up in the harbor. You have to do as much as you can to the point that it doesn't affect other incomes on the island. And you can only do so much without drastically affecting the tourist industry, which is the number one industry on the island. Scalloping - it used to be two hundred and fifty guys did it for five months. Not any more. The fact that Madaket was closed last year, and nobody went in there and took those little growth ring scallops, the abundance of scallops in Madaket this year was as good as I've ever seen it. I dove three times in Madaket this year, and the first time I went I got a bushel of scallops in an area the size of this room. They were stacked on top of each other just like in the old days. And I'd like to see that route gone for a few years and then see what happens, before I go and do all these other drastic things. That, and I think the biggest thing in the harbor is if we built the Jetties back up. I really think it would make a big difference quickly, because the upper harbor does not flow the way it used to. Twenty years ago there was a channel off of Pocomo Point that was fifteen feet deep because of the amount of water that moved in and out of there. Now there's nowhere near the circulation. If you look around here for the last thirty years there's a definite trend - the harbors with the least amount of circulation were the first to lose their scallops, and it slowly cycled across. And for years the Vineyard and Nantucket were the last two to have any amount of scallops. And now we've started to lose our circulation in Madaket and town, and our scalloping has decreased drastically. And then everyone said it's not circulation, because there were no scallops in Madaket. But we proved that that's not true by just not taking that nub scallop. We've got good circulation in Madaket, and we had great scalloping this year. And chances are we'll have great scalloping next year, with the amount of seed that's there. I think the scalloping thing is a combination of several dozen things. It's not one thing where you can go - this is it. It's probably twenty things that changed the scalloping, and probably ten that we can do something about. And what are those ten things?"

March 6

Sunrise: 6:07 Sunset: 5:37
Temps: Low 29° High 45°
Conditions: Calm, clear

fulton fish market

When the proprietors bought Nantucket from Thomas Mayhew, the idea was to form a cattle-growing community. Ten men bought in, and each was allowed to have one partner, so in effect, Nantucket was a commercial enterprise comprised of twenty shares of a corporation. Within a few years, the settlers realized that Nantucket was too small to support a cattle industry, and they switched to raising sheep. Only a few of the proprietors lived or worked on the island, and some immediately sold their shares, much as any investor might. So to perform most of the tasks required to run a small community in 1660, fourteen additional 'half-shares' were added, and several so-called 'half-sharemen' were contacted and hired. The first half-shareman was Peter Folger, whose ability to speak the natives' language was essential, and who had been Thomas Mayhew's longtime right-hand man, and skillful negotiator throughout the whole process. Mayhew and a partner had, in fact, each retained a share at Folger's advice.

Given two half-shares was Richard Gardner, who sailed the *Expectation,* and whose job was to catch and salt cod which swam offshore in the most enormous schools the world had ever seen. Cod could be salted and stored for long periods of time, was a valuable international commodity, and a key component in the ability of the great navies to project their power. Cod fed the navies, and navies, like armies, move on their bellies, to paraphrase Napoleon.

The proprietors had to pay a yearly tax to Governor Lovelace of New York, which governed Cape Cod and the islands at that time, and that tax was in the form of salt cod, which went to support England in its various wars. One such war was against the Dutch for control of portions of the colonies. In fact, the Dutch briefly regained control of their original colony in New York, which they called New Amsterdam, putting the hierarchy of Nantucket's ownership in limbo, and interrupting the Half-Share Revolt. But as long as the Governor was in New York to collect the cod tax, then Richard Gardner delivered it. The fact that he was given two half-shares for his work testifies to its importance.

On a sunny, chilly Monday morning during the first week of March, I drive near Battery Park in New York City, where the Hudson and East Rivers meet up. The Statue of Liberty stands distant in the harbor formed by this giant estuary, as though floating. Skyscraper-sized freighters come and go beneath the cathedral-like arches of the Brooklyn Bridge, recently featured on PBS as one of the world's great architectural wonders. In addition to its architectural importance, the Brooklyn Bridge, when it opened in 1883, linked the ever-growing industrial and residential borough of Brooklyn to Manhattan, forever intertwining their destinies. The bridge abutments, sturdy and massive, lend substance and contrast to the graceful webs of cable and iron they support, and which have kept countless millions of passengers safely aloft in transit since President Chester A. Arthur walked across it to mark its opening. Unlike many utilitarian urban bridges, whose names and designs are unremarkable and easily forgotten, the sight of the Brooklyn Bridge is indelible. This Gothic masterwork is stately and alive, aware of both the real and the symbolic importance of its being, and proudly embracing it.

In the shadow of the Brooklyn Bridge, I look for the Fulton Fish Market. The Fulton Fish Market begins where Fulton Street runs out of land at the lower right hand corner of Manhattan Island. It has been a major center of the fish buying trade in the Northeast since its creation in 1834. I have a mental image of a vast, brick-floored, scale-encrusted square, where scores of men in bloody aprons haul enormous wooden crates, overflowing with whole, flopping fish, off of an ever-arriving fleet of trawlers and draggers, unceasingly slice at them with expert, massive machete strokes, and hurl them into an ever-departing fleet of beaten-up, worn-down delivery vans, which somehow magically distributes them throughout the entire Northeast Corridor, despite the obvious fact that these vans don't look like they could ever make it past city limits.

Instead I find that, by late morning, Fulton Fish Market is closed for the day. Work begins here at 4:00 a.m. And there is no actual fish market named Fulton's. Instead, there are dozens of small storefronts, separated from the vehicle parking lots and truck loading bays by industrial, roll-down, cage-like grates. Most of these operations have concrete floors, several large counters, ice-filled display cases, standing industrial scales, large hanging scales, forklifts, and pallets. The floors are freshly wet from their end-of-day hosing down.

Across the East River I see the old Saint Francis Hotel, once the highest point in Brooklyn, and which still commands a view of the Brooklyn and Manhattan bridges, the graceful curve of two expressways, and the contrasting geometric splendor of the Manhattan skyline, which is so spectacular that it seems generated

by a Hollywood special effects studio. On my side of the river are the many individual components of the Fulton Fish Market: Long Island Seafood Exchange; Louisiana Seafood; Paretti; FairFish; Sunrise Seafood; Messing Fish Co.; Barris Cafe; Joseph Carter Third Generation Seafood; Mount Sinai; Best Buy; Viking; R & M Fish; LaRocca Fish; Jim's Fish; Arrow Fish; Arrow Seafoods; Dino's Seafoods; Lou's Market; Sea Maids; Emerald Seafood; Smitty's; Filet; Pickle Barrel Seafood; Williams; Crown; GNY; Montauk; Gloucester Fish, Lobster, and Shrimp; New Clearwater Fish; Haley's; Universal; Joe Manoni; Buyer Lightning; Blue Ribbon; Lockwood and Winant. The last I recognize as a name that Charlie Sayle sells to.

Once again I am reminded of my first day on the town dock, imagining some vastly romantic fleet sailing off to harvest the scallops, only to understand that a hundred men or so were merely going off to their day's work. I have that same understanding here at the southern tip of Manhattan, as trucks and taxis rush around me pursuing their daily business, while the fish-buyers have already closed up shop for the day and gone home to their families, to congregate here again tomorrow before dawn.

The southern tip of Manhattan, where the great Hudson meets the East River and begins to mingle with the Atlantic ocean. As the new millennium began, the twin towers of the World Trade Center stood as one of the 20th century's defining symbols, just as the Brooklyn Bridge, at the lower right, remains a defining symbol of the 19th.

balducci's

My next stop is Balducci's, a specialty food supermarket which Dave Fronzuto calls Barducci's. He knows Balducci's because a woman who summers on Nantucket calls him every year to report on the day they begin selling Nantucket Bays, and the price. Faithfully, she buys three-quarters of a pound, no more, no less.

Balducci's is located at Sixth Avenue and 10th Street, where Greenwich Village blends into Chelsea, a couple of miles north of Fulton Fish Market. Between them lay artistic SoHo, academic New York University, musical Washington Square, and enough cozy eateries to suit every possible economic, ethnic, social, gender, or culinary preference. This is the spiritual center of East Coast avant-garde chic, and one feels its heart beating day and night.

People in this neighborhood tend to be showy and obsessive about at least one thing, drawing suspicion if they are not. The people who visit Balducci's are obsessive about their food. As I wait, double-parked, for a luxury car to move out of a parking place, a couple opens the trunk and spends at least five full minutes compulsively arranging the handful of items they had just purchased into two ice chests brought for the purpose, before closing the trunk and driving off, entirely oblivious to me, as all New Yorkers seem to be to all cars.

The Sixth Avenue entrance to Balducci's advertises, in elegant gold lettering, poultry and wild game, free range chicken, fresh turkey, squab, duck, venison, and other gourmet items. A large window features a bed of ice upon which is arranged, with the artistry of a Spanish mural, a display of exotic seafood, mostly still whole, such as I never imagined to exist, dominated by a Mississippi alligator gar. To the left of this museum piece, a signboard says: ***Fresh Fish - Today's Seafood Department Features:*** and lists the following items: Alaska Wild King Salmon - 15.99; Florida State Stone Crab Claws - 26.99; Lobster Tail - 37.99; Cooked Shrimp - 19.99. At the top of this list are Nantucket Bay Scallops - 26.99 a pound.

The interior of Balducci's is a surreality of gourmandizement. People pore over an endless variety of exotic, often unrecognizable foods, with tenderness and affection. Having selected, they then, like the couple in the car outside, carefully and gently place them into bags, boxes, or containers brought especially to preserve the look and feel of their goods. Balducci's is no less than a culinary house of worship.

Jacques Adou manages the seafood department as a world-famous chef manages a restaurant. Speaking with a remnant of a West Indian island accent, he tells me that, as far as bay scallops go, he only buys Nantucket Bay Scallops. When they are in season, he sells about fifty to sixty pounds a week. He is out of bays right now - but if he gets some, they go right away. In fact, he has a fifteen pound order from a good customer that he would like to fill, so he is waiting for some to come in. His customers will not settle for anything less than a Nantucket Bay, and they know the difference. Jacques tells me he sometimes gets them from Fulton Fish Market and sometimes directly from a fisherman. But he good-naturedly fails to acknowledge my question as to which fishermen he deals with.

In fact, Jacques says the Nantucket Bay is his own personal favorite, because of its size and sweetness. I ask how he cooks them.

"Very, very simple," he states. "Chopped garlic, salt and pepper, fresh thyme, fresh rosemary, white wine, and butter. Sauté it up for five to eight minutes, no more."

"If a fisherman were to call you up and say that he has some, would you buy them?"

"Oh yeah, right away," says Jacques.

That afternoon I drive to the upper West Side of New York and visit two more well-known specialty stores, Citarella's and Fairway's. Though rivaling Balducci's in many respects, their scallop selection is slim: sea scallops for 11.99 a pound, and some unnamed variety of baby bays for 7.99 a pound. At 26.99 in Balducci's, if available, this gives me my best indication as to how highly prized is the Nantucket Bay Scallop.

peconic bay

Nantucket is now world famous for its bay scallops, but the state of New York was once equally so. In fact, *argopecten irradians irradians* is the Empire State shell, while Massachusetts honors the purple whelk. But New York's Long Island harbors lost most of their scallops during the eel grass blight of 1931, and those that recovered from the initial disaster have periodically suffered damaging, recurring brown tides.

Chief among New York scallop fisheries was the Peconic Bay at the eastern end of Long Island. Peconic Bay Scallops were once so prized that they were also referred to by name, like Nantucket Bay Scallops. The Peconic scallop fishery has provided winter income to many 'baymen', as they are called, and their families, who have fished these waters for three hundred and fifty years. In the past, large harvests of Peconic Bays have actually glutted the market, holding down prices. In 1974, as a matter of fact, Nantucket Bays were selling for $1.60 at the dock, partly because of good harvests in the Peconic. In 1982, more than 500,000 pounds, or roughly 65,000 bushels, of Peconic Bays were landed, numbers equivalent to Nantucket landings.

Two disastrous brown tides of mysterious origin, but thought by many to be caused by nutrient overloading, virtually wiped out the entire Peconic scallop fishery in the mid-1980's. But the bay was soon re-stocked, partly with propagated Nantucket scallops, and by the early 1990's the baymen were once again landing landing significant numbers. Then, in 1995, another brown tide struck. The following season, fifty pounds of scallops were landed in the Peconic, according to the National Marine Fisheries Service. This translates to six and a half bushels, barely a day's limit for a Nantucket scalloper. In other words, for every 10,000 scallops living in the Peconic Bay just fourteen years earlier, there remained just one. If a similar disaster hit the United States population, that means there would be less than 30,000 Americans living. And if a similar disaster struck Nantucket, there would be less than one person left. Talk about your half-shareman.

March 7

Sunrise: 6:04 Sunset: 5:39
Temps: Low 30° High 47°
Conditions: Calm, springlike

the whole package

The last remaining Nantucket man with native blood, Abram Quary, died in 1854. He lived on family land still known as Abram's Point, a highly visible and beautiful reach on the south side of the harbor. Abram was once arrested for firing a rifle over the heads of souvenir hunters on the site. His defense, that he was protecting the last resting place of his ancestors, was deemed worthy. It was also true, since his parents were buried there.

A very noisy controversy erupted again in the last decade regarding the same piece of land. The revelation that an enormous trophy home was planned for the site triggered protests based on the environmental and visual impact of such a large project on the harbor. The 1980's had brought the first round of such protests against huge waterfront homes with the building of an absolute monstrosity in Pocomo, and this one seemed to ignite even more fervent controversy, as no protections were in place to prevent visual pollution of the view from the harbor.

In the early 1970's, a timeshare condominium development in Madaket caught the town unawares, and the town very quickly created a Historic District Commission to review new buildings for historical appropriateness. It was an acknowledgement that Nantucketers understood they had something very special in the architecture here, a symbolic connection to the past, and they knew how easily that could be lost. Over the years the HDC, as it is called, has been able to expand its range and tighten its restrictions. While controversial simply because it exists, (nobody likes to be regulated, although they do appreciate cashing in on a place that is more valuable precisely because it is regulated), the HDC still had no jurisdiction over views from the water, or size. The decade of the 1990's saw a simply overwhelming number of trophy houses being built, encroaching further and further toward the waterfront, with ever larger lawns requiring fertilizer, and symbolically cordoning off the shoreline with shingles and trim rather than beach access and brush.

As I checked the boat basin this morning, I noted three scallop boats docked at slips three, six, and seven. The island has begun to look spring-like. The sky and the water are losing their winter crispness, as the warmth allows moisture in the air to soften the contrasts. The sun is higher in the sky as well, which adds even more to the effect. Fred Conner was out in his boat *Burnt Swamp* for seven hours and caught two boxes. Like others, he would not tell where he fishes or who he sells to. "You build up a catalogue in your head, and you try to keep it from newcomers," he said. Spanky hauled out last Friday for his vacation. He said fishing was pretty good after the deep freeze was over, but he wouldn't reveal his buyers either. He is looking forward to a vacation. When asked about the recent decline in scallops, Spanky tends toward blaming global warming as well. He says the grass is good in the western end, and also that he had good fishing in the Chord of the Bay and off Dionis this year. But he also echoes what a few other scallopers have said: "How are we supposed to know what's causing it, when the biologists don't?"

It may be a beautiful day, but even diehards are beginning to drift toward shore work. In last week's newspaper, it was rumored that the ultra-pricey White Elephant Hotel was paying workers $100.00 an hour in order make sure it would be open by summer. It has been undergoing renovations since being purchased by Nantucket Island Resorts, which is headed by a Massachusetts-based developer and sports-stadium owner who began making inroads into the Nantucket hospitality business in the mid-1980's by buying and renovating the Wauwinet Inn. The White Elephant faces town across Children's Beach and the Yacht Club moorings, and has long been associated with the finest in Nantucket accommodations. Also included in the purchase were the Boat Basin and the Harbor House Hotel and Cottages, the core holdings of a real estate venture originally put together by Walter Beinecke and his company, Sherburne Associates, in the 1960's.

Beinecke had sold his interests to a large investment firm named Winthrop Associates in the mid-1980's, who managed it with mixed results from both a financial and an aesthetic standpoint. And in 1999, Winthrop sold the whole package to the present owner, who made news by being forced to deny that they were paying that hundred dollars an hour. Given the fact that roughly seventy-five tradesmen are being flown in every day to meet the shortage of workers on the island, the rumor was not so implausible.

Vince Vacca also resigned from the Conservation Commission. He cited the fact that the elected board had become too pro-development to carry out its mandate. I heard another story today regarding one of the above-mentioned large harborside trophy homes. The landscaper, who is widely known for her organic and natural approach, had convinced the homeowners to build a catch-basin to trap fertilizer runoff before it could leach into the harbor. Unfortunately, the builder, fearing it might delay the project and cost him money, stymied it, costing the harbor and the scallops another small piece of their habitat. It was yet another illustration that greed is not just in the realm of the wealthy, nor is ignorance the sole property of the uneducated. It was also an illustration of how difficult the task of saving a harbor truly is, when even a well-meaning landscaper and an enlightened homeowner willing to foot the bill cannot accomplish a simple, small protective measure, when a few bucks or a few days' inconvenience get in the way.

Road construction now makes it impossible to get to the Marine Department or Sayle's Seafood. A minor controversy had erupted early in the season when one element of the road paving project, which was simultaneously a sewer line replacement project, involved the replacement of any damaged storm drain pipes that ran under the road, so as to not have to dig up the road again when, at some future date, a storm drain project was initiated. Some harbor protection groups, who are pushing for elimination of storm drain runoff via catch basins or some other future technology, were angered that this project did not address storm drains in the first place and, that by replacing broken or malfunctioning pipes, were actually increasing storm drain runoff into the harbor.

Although the headline of this chapter, the whole package, seems to imply the reference to Nantucket Island Resorts' purchase of the old Sherburne Associates kingdom, in reality it refers to the whole package that needs to be saved if the scallop industry is to be saved: the animal, the industry, the habitat. As Dave Fronzuto had told me, that is what Nantucket uniquely brings to the table - an approach that addresses all three of those areas.

St. Patrick's Day

Sunrise: 5:49 Sunset: 5:49
Temps: Low 30° High 54°
Conditions: Snowing

OZ

SNOW on St. Patrick's Day!! And only four days until the start of spring. And on March 18th it was still snowing and windy as I brought myself and two Irish friends to the Mary P. Walker Auditorium at the high school to see the student performance of *The Wizard of Oz*. There are many young Irish men and women on Nantucket, here on student work visas. They began coming in the mid-1980's as the season grew longer and American summer college workers could no longer fill all the jobs. Many have stayed permanently and begun families. Some have even married Americans and returned to Ireland with them. The lilt of Irish accents hangs pleasantly in the air around town, and a careful listener can even begin to discern which part of Ireland the speaker might be from. And although yesterday was St. Patrick's day, the true Irish, while observant, do not generally exhibit quite the unbridled passion for celebrating it as does the many-generations removed Irish-American. In recent years, Jamaicans have begun to migrate to Nantucket for the work as well, and the peculiar cadence and phrasing of that island nation adds its own flavor to the streets.

Many sons and daughters of scallopers filled *The Wizard of Oz* cast, and a full house, or roughly eight-hundred people, were in the audience for the show, which ran two nights. There is something to be said for a community in which almost twenty percent of the residents will turn out on a cold and snowy weekend to see a high school play.

March 30

Sunrise: 5:27 Sunset: 6:03
Temps: Low 24° High 50°

a healthy harbor first

The Inquirer and Mirror ran their election preview issue today, capsuling all the articles and issues that will be voted on. Candidates and interest groups take out ads urging participation. Rob (who is the third Robert in the Benchley line) had told me, somewhat tongue-in-cheek, that he might use RB_34H_20 as his water commission campaign slogan. It would have ranked with one I did see, for school committee: GET OFF YOUR FANNIE, GO VOTE FOR ANNIE.

In addition to the ads and coverage, the *Inquirer* also runs an occasional editorial. Of all the one hundred and three warrant articles they could have editorialized, they chose Article 55, Steve Bender's boat washing issue, headlining it **A Healthy Harbor First**. The body of the editorial read:

> An article in this year's warrant asks voters to consider banning the practice of allowing boats docked in the harbor to wash their boats with soap.
>
> The argument against this method of cleaning is that it is harmful to the environment - our harbor. The counter argument we have heard is that people have a right to keep their boats clean. Sounds like the American Way to us.
>
> But really, what's more important, a shiny luxury yacht or a healthy harbor? We'll vote for clean waters and happy, healthy scallops over gleaming ships any day.
>
> Nantucket is compromising too often these days when it comes to protecting our natural resources. People seem far too concerned with making a fast buck than acting as responsible stewards of the lands and waters around us.
>
> Town officials say all the studies aren't in on this issue and recommend voting against it. We say, better be safe than sorry. Let the yachtsmen live with a little dirt on their boats this summer while we work at protecting what really counts - the harbor and waters that nourish us in more ways than one.

The election for town officers always comes before town meeting, in this case on April 4. The warrant articles are then voted on separately at town meeting, which will begin this year on April 10. The biggest race is always for Selectman. Steve Bender, sponsor of the boat washing article, is running against perennial public servant Jack Gardner, a direct descendant of the original half-shareman, and Matthew LeDuc, a virtually unknown restaurateur. LeDuc does not figure to be much of a factor, but Gardner is the prohibitive favorite. He has been a selectmen for twenty years, on the planning board for thirty, the planning commission for twenty-five, the water commission for twelve, the traffic commission for fifteen, and the park and recreation commission since its inception twelve years ago, even having the Jetties Beach tennis court and recreation facility named after him.

While some consider Bender on the radical environmental fringe, he brings an interesting resume of his own to the ballot booth. He lists himself as a fisherman, age sixty-five, and a nine year member of SHAB. He has a B.A. in chemistry from Alfred University, Masters in math and physics from Northeastern University, and has done postgraduate study in X-ray crystallography, crystal physics, and chemical engineering. His statement to the newspaper reads: "The Comprehensive Plan is brilliantly written and addresses the problems of Nantucket on the mark."

The big issues on Nantucketers' minds these days are the Comprehensive Plan, handling of the Steamship Authority issue, lack of affordable housing, and overdevelopment, and Bender believes a desire to get rid of the current bad government will carry him. But it seems equally clear to many that a vote for Bender is a vote for a radical approach to cleaning up the harbor. Either way, few people on the street give him much of a chance.

Steve Bender

March 31

Sunrise: 5:26 Sunset: 6:04
Temps: Low 34° High 52°
Conditions: Clear, still

last day

Barrett Pier is empty this morning. Five scallops boats float listlessly on their moorings. There is not even a breath of wind for them to point into. The beach grass, which waved green and glorious against the brisk November breezes, now lies brown and flat. Over at Jackson Point, there are nine boats on their moorings, none fishing. At town pier, the cottage named *Scallop* is on the back of a trailer, moving slowly down still-unpaved Washington Street, off to worthy new owners, one hopes.

In the traditional rhythm of the island, scallop season ends today, and construction season begins. Although truthfully, in the past few years, construction season has never really ended. There is always a spring rush in the trades, as home and business owners realize tourist season is almost on them. Few have figured out that one is far more likely to get work done on time, at a fair price, and at a higher level of quality by getting it done in the fall. And chances are they have more money to spend following the summer season, and that whatever needs doing has to get done one way or the other before spring. Tradesmen take advantage of the laws of supply and demand come spring, when deadlines loom and jobs outnumber workers by a large margin. A guy told me someone had given him an estimate of ten-thousand dollars to paint his garage, a job that was worth about two. Hell, I'd have done it for nine.

clarity

A Secchi disk is a marvel of simplicity and elegance used to measure water clarity. It is a white saucer, twenty centimeters in diameter, which is lowered into water until it disappears. This yields a measurement known as Secchi disk depth, which scientifically validates common sense: the farther one can see through water, the clearer it is. It is one of the few methodologies that satisfies both scientist and fisherman. Whatever else Mr. Secchi may have accomplished, that alone is a career maker.

Tracy Curley had told me that the water is at its clearest in early spring. This last day of March bears her out. On our final photo journey, Rob and I can see straight down to the bottom from virtually every vantage point, observing little sand rows created by tidal currents, isolated patches of eel grass waving lazily in the ebb and flow, silvery schools of small fish darting about, various bottom dwellers searching for food. From the ferry or an airplane, travelers can see the contours of the harbor and surrounding shoals, where rips swirl sand out to sea in elongated commas, and slowly fill in lees on the north side of the island. One can see the same effect inside the harbor at the tips of Coatue's points. Dark patches show where eel grass meadows are strong and healthy, and light patches show where the bottom is barren.

Nantucket Harbor was studied in 1998 by Northeast Aquatic Research LLC, a study commissioned by Nantucket's Marine and Coastal Resource Department. It followed on water quality monitoring studies by Woods Hole Oceanographic Institute, also commissioned by the town, which began in 1991 and concluded in 1994. With continuous monitoring and data collection in the intervening years, the report was able to compare data existing in 1991 with currently available data. Both studies measured turbidity; developed temperature, oxygen, and salinity profiles; water quality constituents; and plankton population estimates. The second study was released on September 15, 1999, two weeks before the recreational scallop season opened. The original Woods Hole study concluded that, while water quality in the harbor was relatively high, nutrient levels exceeded those in Nantucket Sound at certain times, and were believed to be coming from the harbor drainage basin and harbor sediments.

Suspended particulate matter in water causes turbidity, (which is technically what the Secchi disk measures), and is an indicator of microscopic life. This microscopic life tends to consist of phytoplankton, or in other words, algae cells. Scientists use Secchi depth observations to estimate phytoplankton populations. The type of phytoplanktons present affect what kind of life the body of water can support.

Jackson Point

Barrett Pier

Secchi disk depths in Nantucket Harbor fluctuate from 6 to 8 foot minimums in late winter and again in mid-summer, to 18 to 22 foot maximums in early winter and late spring. Approximately ten percent of incident surface light reaches Secchi disk depth, and this ten percent barrier also marks the limit where eel grass can grow. In other words, the murkier the water, the shallower it needs to be to support eel grass, and therefore scallops. The two studies showed significant lengthening and intensity of the summer Secchi minimums over the past five years. The 1994 Woods Hole study estimated that the harbor could support eel grass meadows to a depth of about 12 feet. The 1999 Northeast Aquatic Research study suggests that, only five years later, the eel grass limit had risen to a depth of 8 to 10 feet. Given the amount of harbor territory that falls into the 8 to 12 foot depth, an enormous amount of scallop habitat was lost in that five year span. That same time span exactly correlates to the drastic downturn of the fishery.

In addition to the problems caused to the eel grass by this extra turbidity, the types of phytoplanktons causing it affect the harbor. The winter minimums are caused by beneficial benthic diatoms, which provide scallops with their food source for the process of rejuvenation, regrowth, and spawning during the spring ahead. The summer minimums are caused by less beneficial dinoflagellates resulting from increased nutrient levels. The brown tides that periodically kill off the eel grass in Long Island Sound and the Peconic Bay are thought to be caused by algae blooms associated with similar increased nutrient levels.

Finally, a measurement related to clarity is 'compensation depth'. Compensation depth is at two and a half times Secchi disk depth, and is the level below which no primary organic production takes place, due to oxygen depletion. Existing plant life decomposes without regrowth. When this happens in a normally healthy area, it is called an 'anoxic event', and is an early warning signal of trouble. On opening day, Tracy had taken me to a map of the harbor and showed me the Head of the Harbor and Quaise, the two deepest areas. During the previous summer, both had experienced anoxic events. If the calculations of the study were accurate, this occurred at the sixteen foot depth level, which would have covered over 500 acres. And by Belding's estimate, there are only 3,000 fishable acres in the harbor to start with.

Carl Sjolund's shanty awaits next season.

April 5

connection

On Lincoln Circle, two stately houses of Queen Anne shape but with New England austerity, lacking the Victorian-era detailing, long presented side-by-side sentinels overlooking the approach to the harbor. Their tall, narrow silhouettes on the bluff were reminiscent of a scene from *Wuthering Heights.* Since I last traveled this way, one has been skinned and disemboweled, and wind whistles cleanly through the skeleton.

There is a company in Somerville, Massachusetts, which rescues lumber from old barns and mill buildings, planes it, and sells it at no small premium to homeowners for flooring. They say their customers like the feeling of a connection to history. While this company and its supposedly new concept is so successful it made the evening news, Nantucket carpenters have been rescuing lumber for as long as anyone can remember, beginning with the tradition of salvaging shipwrecks, and continuing through today's practice of rescuing materials from the dump, or from demolitions such as the one on Lincoln Circle. My office is in a loft above my living room, and rests on floor joists removed from the Atheneum library during its renovation. I like to brag that Frederick Douglass gave his first public speech while standing on them. Technically, it was not his first speech, but rather his first before a mixed white and black audience. And actually, he gave the speech before the original Atheneum burned down in 1846, and my joists came from the rebuilt structure. But, it is as easy to romanticize history as it is the sea, and who is to say that a symbolic connection is not often, after all, as real as a physical connection?

People do like to feel connected to the world. A completely manufactured home strips away one layer of that connection, which makes me wonder why someone would buy an old house on an old island for an inordinate amount of money, and then spend an even more inordinate amount of money making it look like a new house.

True, a new era has dawned, one which will cocoon us safely with many layers of insulation. But much life remains to be lived in this era, and we are losing that connection to nature. Perhaps an endeavor like scalloping, a throwback to that natural era, is our symbolic connection with it, and perhaps that is more important than we think. Because we are, for good reason in certain cases, and for convenience in others, severing our ties with nature. But whatever the reason may be, something about this particular industry, in which people pull their living directly from the sea, makes it closer to the natural order of affairs.

And perhaps that is part of the reason for yesterday's surprising election result. Steve Bender won by a landslide, 1,330 votes to Jack Gardner's 994. Cantankerous environmental radical or not, the voters seemed to have sent a clear signal that they want a man who wants a clean harbor. What remains to be seen is how the boat washing issue will do at town meeting.

April 15

town meeting

The state periodically sends a note to the town reading like this:

> To the Constables of the Town of Nantucket:
>
> Greeting:
>
> In the name of the Commonwealth of Massachusetts, you are hereby directed to notify and warn the inhabitants of the Town of Nantucket qualified to vote in Town affairs to go to the Nantucket High School at 10 Surfside Road in said Nantucket, on TUESDAY THE FOURTH DAY OF APRIL, 2000, BETWEEN THE HOURS OF 7:00 A.M. AND 8:00 P.M. for the following purpose: to cast their votes in the Annual Town Election for the following offices:

It then goes on to list the offices up for election. There are twelve elected posts, governing bodies, or advisory boards one can be elected to in Nantucket: Moderator of Town Meeting; Selectmen, who function as collective mayor; Historic District Commission and Associate, who determine appropriate building styles; School Committee; Housing Authority; Land Bank; Planning Board; Water Commission, (one for Nantucket and a separate one for 'Sconset, which is the one Gerry Eldridge conned Rob into running for); The Steamship Authority; and the Harbor and Shellfish Advisory Board (SHAB). The Commonwealth goes on to advise:

> In addition, you are directed to notify and warn the inhabitants of the Town of Nantucket qualified to vote in Town affairs, to meet and assemble themselves at the Nantucket High School Auditorium at 10 Surfside Road in said Nantucket, Monday, April 10, 2000 at 7:00 p.m., then and there to act on the articles within the enclosed warrant.

Most of the 103 warrant articles deal with the mundane nuts and bolts of running a town, and are passed with little or no discussion. For instance, the first 19 articles are submitted by the Board of Selectmen themselves, and deal

generally with transferring funds from one or the other budget or fund to the appropriate fund or expenditure for ongoing public works. Article 7 regards the employment compensation schedule of the various town jobs, including a notable change to the shellfish inspector salary from $175.00 per week for twenty weeks to $35.00 per day to be paid when scalloping activity is taking place.

There are twenty zoning by-law amendments, ranging in importance from the definition of a studio to the creation of three new districts in 'Sconset, to the adoption of a building cap to slow development. The rest of the mostly citizen-sponsored warrant articles run the full gamut of human experience. Article 26, submitted by John Doody et al, subtitled Appropriation: Drug Sniffing Dog, states:

> If illegal drugs and their human toll can be eliminated anywhere, it should be possible on an island. This article is to see if the Town will: Appropriate the sum necessary to purchase, house, and maintain a drug sniffing dog all year round on Nantucket, along with its necessary human staff and mechanical support equipment. It is suggested the Police Chief include in the dog and Officer's duties the task of patrolling the airport and docks, riding the ferries, and generally being conspicuous.

Other articles range from a call for a limit to the number of days that a party tent can be placed on a residential property (nine), to a variety of town-sponsored eminent domain takings of beach access routes. In opposition, three related citizen-sponsored articles, 23, 24, and 25, seek to either veto outright or prohibitively limit funding for the takings.

Also included are seven warrant articles relating to the health of the harbor. Articles 29 and 30 prohibit new construction of piers, docks, and wharves except in the town's Residential Commercial District, or by government entities. Article 47, submitted by Phil Austin, sought to radically amend and strengthen the town's by-law, which was essentially the product of the creation of the Harbor Watershed Protection District, which had taken three years of legislative wrangling to craft. Article 54 proposed to ban jet skis from the harbor, and Article 56 sought to exempt rescue personnel from that ban. Article 57 allowed water dependent business to operate from town owned docks, wharves, and piers. But by far the most controversial was Article 55, Steve Bender's proposed ban on boat washing.

When all the giving and taking, amending and tabling, shouting and shushing, voting and counting was over, the shellfish officer's compensation changes were passed, the drug sniffing dog was nixed, the party tent limitation passed, most of the beach access eminent domain takings passed, the article to veto them failed, and the two to limit them were tabled.

Of the articles concerning the harbor, 29 and 30 banning new construction passed, and a watered-down version of Article 47, the by-law amendment, also passed. Jet skis were banned except for emergency personnel use. The boat washing bill didn't really pass, but it didn't totally fail, either. As amended by Bobby DeCosta, the bill did not prohibit wash downs, but limited the types of cleaners to non-toxic, biodegradable, environmentally safe products. As a small step, the measure succeeded since it was the first prohibition of any kind against boat washing in the harbor. But as Bender said afterwards, the toxicity of the soap itself is not the issue. "The most toxic substance to shellfish is the incomplete combustion of fuels. If you leave them alone on top of the water, they evaporate. If you introduce detergents into the water, they enter the water column, impact the shellfish, and the shellfish are dead."

I recall my earlier conversation with Bobby DeCosta and Tracy about the boat washing article, and about Nantucket's status as a water-based tourist community. The unanswered question then, as now, seemed to remain: What, after all, *is* important? And what are real needs as separated from assumed needs?

Nantucket may well be sheltered from the vagaries of the real world by money attached to the giant corporations which provide the wealth that distinguishes Nantucket from any one of a hundred geographically similar islands off the American coastline. Homage is paid at every town meeting. 'We need the tourist dollar' is chanted like a mantra, an obligatory preface to any anti-development proposal, until, by sheer repetitive force, it takes on the mantle of fact. Yet, people have lived here for hundreds of years, sometimes with, sometimes without, the benefit of great wealth. Thousands of communities the size of Nantucket survive and even flourish without the benefit of being a destination for the wealthy.

But the population of Nantucket, like the world at large, grows daily, and the supply of non-renewable resources shrinks daily. Every person born uses some of them. Economic, political, and psychological reasons, reasons which are very deeply rooted in our survival instinct, make it easier to ignore than to face the inevitable conclusion that this knowledge points to: the upward trend of population must eventually intersect the downward trend of resource. Sooner or later, we will run out.

To compound matters, the laws of supply and demand tell us that as a given resource vanishes it becomes more valuable. Therefore those that profit most by harvesting it become richer, more influential, and, temporarily at least, greater beneficiaries of their own short-sightedness. They cannot be expected to have much self-interest in its conservation - after all, they are being paid handsomely to remove it. By the same

token, neither do they deserve credibility or control over management either, due to the obvious conflict of interest. Which, in a nutshell, is why resource management in a democratic society becomes the business of all the people, rather than merely the people who profit from it. Whether that resource is scallops or oil.

Yet, before we judge too harshly the corporate powerbrokers who make Nantucket their vacation home, and therefore contribute their share to its declining scallop industry, we must acknowledge that the world has long since passed the point where an agrarian economy could supply its needs. Without agribusiness, vastly more people would starve. Without mass production, only the extremely wealthy would be driving cars, using computers, getting open heart surgery. Many residents or visitors who run these corporations are also friends, employers, and benefactors of Nantucket's remarkable array of cultural and social activities, and many islanders might miss them for reasons other than their money.

Remember the survival quatrain: sustenance, safety, resistance, procreation. Agribusiness, with all of its fertilizers, pesticides, hormones, and machinery, is our food source; enormous military machines, heavily armed police forces, and millions of prison cells keep us safe from our only genuine predators, ourselves; and the pharma-medical industry has rendered disease a personal tragedy to those unfortunate enough to be afflicted by it, but more or less an afterthought to the rest of us. These survival tactics are effective and necessary, they benefit all of us, and the impact if they were suddenly removed would be disastrous.

Which is a long way from saying that there is no longer a place for simple, traditional, individual virtues and quirks to flourish. Thomas Jefferson wrote *"I study war and politics, that my children might study business and commerce, that their children might study art and music and literature."* He understood the order of the world, the essence of mankind: the need, unique among creatures, to go beyond the survival quatrain, to reach a point where the challenge is not in discovering how to survive, but in discovering why.

Maybe the spoils of the corporate survival machine indeed make it possible for Nantucket to survive as such an individualistic society. Or maybe it survives in spite of them. In the meantime, twenty-five hundred boats need washing all summer long. Perhaps their owners supplied ten thousand people with food last week, or saved three hundred lives with a vaccination, or perhaps negotiated a truce in a small civil war. Perhaps seeing their boat newly washed gives them a significant amount of pleasure. Perhaps it adds to their reason for being.

And if that means that next winter one scalloper catches only thirty pounds of scallops a day instead of forty, one might be tempted to say 'so what?' Nantucket Bay Scallops, one might also contend, are not exactly the staple diet of the starving masses. If it seems hard to put a dollar value on the pleasure a boat owner derives from a clean boat, that is because there is no real way to measure it. But it is easy to put a dollar value on a scallop. As Kenneth Holdgate told his daughter Linda when she was a little girl, a typical Nantucket Bay Scallop is probably worth about a quarter.

But how much would the last remaining Nantucket Bay Scallop be worth?

May 18

weariness

The business of the island rolls on long after scalloping ends. The first quarter real estate figures were released and compared with 1993. Seven years earlier, $30,238,000 worth of real estate sold in the first quarter. In 2000, that number was up over five and a half times, to $170,299,000. The average price of a house jumped from $344,000 to $1,205,000, with a median of $600,000. By comparison, the most expensive metropolitan area in the country, the San Francisco Bay Area, had a median price of $369,700. If a scalloper owned an average house in 1993 and sold it in 2000, he'd have made the equivalent of a lifetime of scalloping. Or, put another way, if he had a lifetime of scalloping profits suddenly dropped into his pocket, he'd still be three or four hundred grand short of owning an average house.

A very distressing front page story headlined the May 18 issue of *The Inquirer and Mirror*: **Marine Dept. May Lose Biologist**.

Tracy asked for a three month leave of absence. “I need to recharge my batteries,” she was quoted as saying, and intends to spend more time with her son Dana, and to finish a children’s book she is writing about piping plovers. The reporter describes her as growing more weary by the minute. She has been on the job for seven years now, the same span of time those real estate sales figures measured. During that time the homes and boats have increased, and the scallops have decreased. And along with the scallops have decreased many of the other reasons for living on Nantucket that are perhaps best symbolized by scallops. Many people have noticed and become depressed at this correlation. In acknowledging the moment of weariness described by the reporter, Tracy has lost none of her straightforward eloquence: “The reason people want to come here is because of the beautiful environment that we are slowly destroying. The changes are so gradual, but the end result can be catastrophic.”

Epilogue

Summer, 2000

November 1, 2000

ashes

Thirty five mile-an-hour winds greeted the fleet on opening day of the 2000-01 scallop season. The price was $13.00 a pound again. Most of the thirty or so scallopers who fought the weather caught their limits. John Backus did not. He was busy going from fishing spot to fishing spot, spreading his father's ashes in all of his favorite places. "There weren't too many places he didn't know about," he told the newspaper. Eddie Backus was another of Nantucket's old-time scallopers, in fact one whom Tracy had fished with ten years earlier. He had died over the summer, after forty-five years of scalloping. He had been quite ill most of last season, but I had heard that he was fishing some, with a little help from John.

Events that began unfolding late last season were now coming to a head. Rumors had swirled in September that areas might be off-limits to family scallopers. They proved true, and most of Madaket was closed to family scallopers due to high levels of seed. In town, it took several hours to get less than half a bushel. The scarcity of scallops did not bode well for anyone.

In mid-October, Dave Fronzuto announced that Mike Hickey was coming down from the state, that test dredges would be taken from various areas around the harbor, and that if there was too high a ratio of seed to adult, commercial areas would be closed as well. At a tense meeting on October 26, attended by roughly fifty scallopers, Dave announced that a 50/50 ratio of seed was found at most locations, which was, in his view, enough to keep all areas open. But he was not backing down on his determination to enforce the growth ring statute as it reads in the state law. The January meeting with Dave Leavitt of Woods Hole Oceanographic Institute had convinced him once and for all that the 'nub' scallop will live to spat again. Dave has also lobbied for fewer inspection points, and has indicated that he will be touring the shanties more often.

Importantly, Mike Hickey spoke at the meeting and stated that he had been wrong in 1997 when he told scallopers that nub scallops exhibiting many signs of age, such as thickness, coloring, or barnacles, were adults. "I made a judgement call in 1997 that I will stand by," he told the hostile crowd, in that same mellifluous voice that had played a practical joke on Dave Fronzuto a year earlier. "I'm sitting here before you saying, looking at the information now at hand, it wasn't the best call." He also pointed out that fisheries have disappeared up and down the coast. "They disappeared from New Bedford and it's creeping down the Cape and guess what? There are lots of pressures on the scallop habitat, and anything you can do to help the industry is money in your pockets."

Blair Perkins' view, as recorded in *The Inquirer and Mirror*, was this: "The town is going to have to change. Enforcement needs to be more visual. We can't cut each of them open." Dave Fronzuto would reply that Blair is exactly right, that you can't cut each one open, and that the 'well-defined raised annual growth line' criteria in existence since 1909 still works because it makes enforcement about as visual as you can get: "If you have to look, throw it back," he would say, borrowing Tracy's phrase, who did, in fact, return to her post.

Also on opening day, a newly hired warden was very quickly welcomed to the Nantucket scallop fishery and the growth ring controversy. After he inspected and rejected a scalloper's first day catch, the irate scalloper pushed him, threw a punch at him (which missed), and attempted to run him down with his pick-up truck. For which he was given a day off of fishing and a court date.

abalone

Scientists in California were recently thrilled. After two years of trying, they had finally collected fifteen white abalone.

White abalone is the most prized of all abalone, coveted for its sweet meat and exquisite mother-of-pearl shell. It is sort of like the Nantucket Bay Scallop of the West Coast, only with better housing. White abalone inhabit the deeper areas of the sea floor from Point Concepcion to Baja California, once paving them in numbers ranging from two to three thousand per acre. Given their much larger size, I believe this would qualify for the Nantucket term 'ass-deep'.

A 1979 University of California study said that the white abalone, being broadcast spawners (like scallops), would never grow extinct. Almost immediately after this pronouncement, they were commercially fished out, and twenty years later it took the California scientists two years to collect their fifteen in an effort to breed and save them from the extinction they had originally deemed impossible. The abalone in the wild are now typically tens to hundreds of feet apart. That's a lot of sea water to expect a tiny egg and a tiny sperm to hook up in. Even in California.

Dave Fronzuto estimates that 14,000 bushels of scallops were landed in 1999-2000, a very heartening turnaround. On the other hand, 14,000 bushels is only ten or fifteen percent of what Nantucket, with over two hundred boats fishing for five months, consistently sustained thirty years ago, and only a third of a typical catch just ten years ago. It must also be remembered that almost every other northern bay scallop fishery has, at one time or another, closed entirely, with the exception of Martha's Vineyard, which puts 5,000,000 seed a year into their harbor.

While landing data before 1980 was more guesswork than anything, it is still probably safe to say that Nantucket's catch regularly exceeded 100,000 bushels. 1968 likely saw upwards of 150,000. A brief history of the past two decades shows how the scallop fishery has changed radically, in direct correlation to Nantucket's population growth.

From the mid-1970's until 1997, most of the important events in the scallop industry involved market forces or natural disasters. What rules were made were more-or-less fine tuning of long-accepted procedures. For instance, in 1977 the limit was six full burlap bags, which by some accounts was like bringing in nine bushels. But in 1978, bushel boxes replaced burlap bags, which reduced the scalloper's daily catch, and therefore his income, by roughly one-third. During the first two years of bushel boxes, Nantucket landed 59,000 and 96,000 bushels respectively.

In 1980, 400 family licenses were issued and areas of Pocomo and Shimmo were set aside for family fishing only. 379 commercial permits were granted and 130 boats fished all winter for scallops which sold for $6.25 a pound, a relatively high price probably due to the entrance of a fourth buyer named George Harrington. A year earlier, Demo Marken had tried to enter the market, and the price rose to $4.65, but it fell back again to $3.65 as soon as he dropped out. The average piece of land was about $30,000 that year - the same year that 117,000 bushels were landed.

Yet, on opening day of 1981, shellfish warden Allen Holdgate predicted the season would be way off, telling *The Inquirer and Mirror* - "The old time scalloper who wants to put a full day's work in can hack out a living, but the guy who's used to having them practically jump into the boat, like they have in the past, is gonna have a hard time." In Tuckernuck, scallopers complained of too much eel grass. To add to the woes, frozen scallops from

1980's record crop, and the entrance of the southern bay scallop depressed the market to $3.25 a pound. And worries surfaced that the Madaket fishing grounds were shoaling up due to sand sifting through the opening between Smith's Point and Esther's Island. In this 'off year', 77,900 bushels were caught.

The five box limit was instituted in 1982, and scallops opened at $4.25 a pound. Only 304 people got licenses and only 104 boats fished. Still, 50,000 bushels were landed. 294 people got their licenses in 1983 and another new rule was added: no fishing below 28 degrees. It was the birth of the red flag.

In 1984 a planned dredging of the Madaket channel caused the unlikely pairing of scallopers and the Conservation Commission to fight both the timing (October) and location, because both saw the clear danger of silting up the harbor just as the season opened. The project was both delayed and the proposed new channel moved. Otherwise, the years 1982 through 1985 were uneventful, averaging 43,000 bushels. Landing data for 1986 is lost, but likely fell within that range. The fishing effort bottomed at 243 licenses issued that year (curiously, one third of these permits went to women), probably because the first building boom had struck, and there was plenty of high-paying work on land. 1987 and 1988 brought the two worst landing years since accurate records were kept, 23,000 and 25,000 bushels, but that is probably explainable by continuing low license totals: people were still banging nails and swinging paint brushes instead of fishing. But George Metcalfe, who had been scalloping for nine years, noted that 1988 was the slowest start ever. And town officials, for the first time, mentioned rain, pesticides, or sewage as possible causes for the shortfall.

Then an unusual thing happened. In 1989, the number of licenses issued jumped from 252 to 440. A recession had started, building had stopped, and people went back on the water. And as the landings prove, scallops were still there, 48,000 bushels of them - well within the expected parameters of a natural cycle. The only downside was the opening day price: $5.00 a pound. New sources had impacted the market. Digby scallops from Nova Scotia, Baja scallops from Mexico, and Calicos from Florida were all competing. This normally might not have made a difference, because, as Boston's Roger Berkowitz, who buys thousands of pounds of Nantucket Bay Scallops a week for his five Legal Seafoods restaurants, put it: "These are the best in the United States." But he added that, in an economic downturn, people will turn to cheaper substitutes.

But by now, Health Inspector Richard Ray was beginning to voice doubts that the poor catches of 1987 and 1988 were due exclusively to natural cycles. "There's no doubt that water quality is playing a significant role in the decrease of the scallops," he said. "Portions of the harbor are teetering on inactivity." He named boats, drainage pollution, and waterfowl as possible causes, and planned to ask the town for $250,000 to fund a Woods Hole Oceanographic Institute study of the water quality in the harbor.

1990 saw another good year, or at least what was getting to be thought of as a good year. Gibby Nickerson told the newspaper: "When they're thick like this, any damn fool can catch them. You could get them if your dredges were upside down." But one man's good year can apparently be another man's bad year, as evidenced by one scalloper calling it quits after thirty-six years due to scallops being too hard to find. Nonetheless, 44,000 bushels of scallops were landed.

Then, what many believe may have been a major turning point happened. Hurricane Bob hit the island in August, rearranging some of the shoals and sandbars. By itself, this may have been an insignificant event. But two months later, on the eve of opening day, the biggest disaster since the Great Fire struck. The No-Name storm, also known as the Halloween storm, pounded the entire Eastern Seaboard, and racked up $30,000,000 worth of damage on Nantucket. Hundreds of thousands of seed scallops washed ashore, most of Brant Point was under two feet of water, houses were destroyed, boats were sunk, decks floated from one part of the island to another, thousands of square feet of shoreline were eroded, and many of the eel grass beds shoaled over. Communities up and down the East Coast reeled in the wake of what eventually became known as 'The Perfect Storm.' Gloucester suffered the most, as six fishermen died aboard the *Andrea Gail.* Nantucket's scallop harvest dropped to 27,000 bushels that year, and the opening price was still low at $5.00 a pound. The only benefit was that many idled tradesmen were put back to work repairing storm damage.

1992 was similar to 1991, with scallops opening at $6.00 a pound and a catch of 24,000. But in 1993, the take dropped sharply, to 13,000. It was by far the lowest total ever seen. John Sudarksy happened to corner the market by offering $7.00 a pound when everyone else opened at $6.00. Alarm bells began to ring, and the Marine Department began its first experimental seeding of the harbor, contracting with aquaculturist Rob Garrison, who had teamed with town biologist Joe Grochowski to form an operation called Nantucket Research and Education Foundation, to plant between twenty five and thirty thousand seed scallops.

The 1994 season was nothing short of bizarre. On opening day, thirty scallop boats had to be lowered into the water by crane because construction on the Children's Beach boat ramp was not quite complete. "Somebody should be court-martialed for this," spat Jim Crecca to *The Inquirer and Mirror*. The price opened at $9.00 a pound, a move Sudarsky claimed was a concerted effort to drive him out of business for cornering last year's market. "They're trying to crush me, but hey, more power to the fishermen," he declared. Bill Sandole of East Coast Provisions and Seafood, an off-island buyer for fourteen years, stood pat. "I don't want to get in the middle of this outrageous price war," he said. The dock price quickly fell to $7.00, then $6.00, then settled at $5.50.

Inexplicably, Martha's Vineyard scallops had been fetching $8.50 a pound since their season opened on October 15, yet Peconic Bays, finally back on the market after ten years, had started selling at $4.50 in early October, and were at $3.00 now. "There's something strange going on," said Fronzuto.

Even more oddly, there was a huge spike in Nantucket's production: 28,000 bushels were landed. No one knew then that it would be the last time Nantucket would land even half that many.

pandora's box

And no one yet realized what controversies over how to save the scallops were in store, or even whether they needed saving at all. But Pandora had begun to peek out of her box four months earlier. A concern voiced by scalloper and would-be aquaculturist Steve Scannell about predation by non-native green crabs prompted the Selectmen to ask Tracy Curley, who had recently replaced Joe Grochowski as town biologist, to look into it. On July 11, after checking with Woods Hole and the Division of Marine Fisheries, Tracy responded that green crabs, blue crabs, and conch are important predators, but in terms of whether a policy to eliminate them would pay off, she noted that the law of nature generally provides that when one predator is eliminated, another fills the niche, so no single predator could be held responsible for the problem. "There is a natural ratio of predators to prey," she wrote.

She went on to say that habitat degradation proved the greatest risk. But she also added - "Due to current fishing practices, the scallops are allowed only one spawn. Hence, propagation is limited. Accidental taking of seed by fishermen coupled with scallop strandings caused by storm events also reduces overall size of the scallop population." Her recommendation was more fishery management, which she outlined as initiating a plan for water quality management, discouraging the taking of seed scallops, habitat enhancement, rotating fishing beds for habitat recovery, and seeding beds. "Unless all facets of the fisheries are considered, there is little hope for the scallop industry to recover," she concluded. It seemed innocuous enough, at the time.

By 1995, 1300 family licenses were being sold, as opposed to the 400 of 1980, when there were four times as many scallops to catch. Private homeowners were also beginning to barricade traditional access routes to the harbor. "Inherently, it will get worse as private owners continue to exercise their right to close off access," said Fronzuto. But efforts to deal with the dwindling scallops, fed by the first scare of 1993 and the knowledge that other fisheries had lost all their scallops, led the town to order 2,000,000 scallop seed from Rob Garrison's NREF. As knowledge and experience accrued, it became evident that only 500,000 could be delivered at the end of October. The parameters had changed from an order of 2,000,000 scallop seed of three millimeters diameter to 1,000,000 seed of twenty-five millimeters diameter. The town had spent $71,422 for seed and improvements to a facility near the Coast Guard Station at Brant Point for NREF to grow them. And as part of an effort to track and understand where the scallops went after they were planted, and how they interacted with the natural set, NREF was attempting to breed a striped scallop. But, NREF's failure to deliver the promised modified total amount set off recriminations with the Selectmen and SHAB, which threatened to acquire future seed from other sources. Garrison argued that NREF should be treated as a research facility and not as a business, because in order to be self-sustaining, a nursery operation had to sell 50 to 60 million seed, and his facility could at best grow one-quarter that amount. But he felt that subsidizing it was worthwhile, because it kept people involved in marine science on Nantucket where they are needed.

In 1996, concern truly began to shift away from issues like the price of scallops, or safety and convenience in inspections, to the recognition that the very existence of the Nantucket Bay Scallop fishery might be in jeopardy. It was just sinking in that the Long Island harbors that had worked so hard to bring their scallops back since the mid-1980's had lost them again. And Nantucket's 1995 and 1996 harvests had dropped to 13,391 and 11,110 respectively. In fact, it was becoming clear that this last catastrophic year had probably ended northern bay scalloping everywhere except Nantucket and the Vineyard. Bay scallop landings in the United States became virtually non-existent except in North Carolina, which catches the southern bay scallop.

In response to these concerns, SHAB asked Tracy to study eel grass mitigation, or the concept of replanting eel grass. In a March 18, 1996 letter, referring to a 1984 study of Atlantic coast eel grass meadow ecology, Tracy wrote that there was insufficient evidence that mitigated eel grass meadows performed the same function in nature as natural ones, and therefore should be used only as a last resort. "Mitigation is a short term solution to a long term problem. If eelgrass is declining, the newly planted eelgrass will decline as well," she wrote. "The Town needs to face the real issues of why the eelgrass is declining and habitat is lost."

She listed four major areas of impact on eel grass:

> 1) dredging and filling
> 2) commercial fishery harvest techniques and recreational vehicles
> 3) modification of normal temperatures and salinity regimes
> 4) addition of organic and inorganic chemical wastes

And she proposed four possible courses of action:

> 1) Reduce entrance of landscape based pesticides, herbicides, and fertilizers into the harbor
> 2) Rotate fishing grounds and limit 'soft sediment' bottoms to divers only
> 3) Investigate more eelgrass-friendly dredge designs
> 4) Have a sliding scale fee for scallopers based on method of harvest, rewarding the least detrimental methods

Pandora was now peering out from the dark recess of her half-open box, just as a scallop peers, blue-eyed, from its half-open shell.

recent history

In October of 1996, as the Peconic baymen were just realizing that they had yet again lost their fishery, newly elected SHAB chairman Tobey Leske went before the Board of Selectmen and made what was described by *The Inquirer and Mirror* as an 'impassioned plea' on behalf of ours: "Nantucket's scallop fishery is not in a state of decline, nor just in trouble. It is in a state of emergency."

SHAB was not asking much - merely to reduce recreational scalloping from seven days a week to five, as they had become increasingly convinced that family scallopers were now taking twenty-five to thirty percent of the total catch, as opposed to ten percent in years past, and that a significant number were taking two to three times their weekly limit of one bushel. None of which had been a problem in the past, when there were many scallops and few family scallopers. But the permits granted had jumped from four hundred to over eighteen hundred in the last fifteen years. Jack Gardner suggested lowering the number of permits granted, while Tim Soverino noted that eighteen hundred family scallopers represented a lot of votes. Pam Killen and Vince Vacca favored the SHAB measure, Vacca saying he'd rather raise the ire of the recreational scalloper than fail to protect the harbor. The measure lost, three votes to two. Tobey told the newspaper he understood the pressure they were under, but wished they had listened anyway. "Our function is to recommend to them changes in the regulations," he said. "That's what we're here to do. They're supposed to look to us on this type of thing."

A month later, the Selectmen were singing a different tune. The first few days of the commercial season showed that an even worse season might be shaping up. They now also knew that the Peconic had lost their scallops, that the Cape was barely hanging on, and that even the southern bay and calico fisheries were in serious decline. Gone many years now were the great scallop fisheries of Long Island, of the Chesapeake, of Barnegat Bay in New Jersey, the Niantic in Connecticut, the Acushnet River and Buzzard's Bay, and both shores of Cape Cod. At best, some towns were catching scallops during some years, but Martha's Vineyard and Nantucket were now the last significant northern bay scallop fisheries, and the Vineyard was putting up to five million seed into their waters every year to keep theirs alive.

In light of this, SHAB returned to the Selectmen with another plan, far more radical than the first - this time to end the season for both commercial and recreational scallopers on January 1, on the theory that the vast majority of scallops are out of the water by then, and that there is no sense in tearing up the eel grass by dragging over it again and again for fewer and fewer scallops. While the Selectmen did not pass this proposal, they did ask SHAB to work with them over the next month to develop a plan to save the scallops. SHAB followed up by using its next meeting to invite scallopers to participate. Among the ideas discussed that evening were elimination of off-island family permits, heavier fines for family scalloping infractions, reducing commercial catch limits, putting a moratorium on dredging, and limiting each boat to five bushels. Tobey Leske said that four to five-hundred recreational scallopers walking on the eel grass cannot possibly be good for the seed. Rob Garrison said no one had studied the effects of dredging for scallops or recreational scallopers walking on the seed. And Steve Bender suggested lowering the daily limit to extend the season, cordoning off areas of high seed concentration, and stressed the need to protect the water quality. Dave Fronzuto was typically blunt in his appreciation of the new-found spirit of cooperation between the two town boards, telling *The Inquirer and Mirror*: "The scallop industry as we know it is in serious danger. It looks like the Selectmen are willing to work with us finally, and that hasn't always been the case. We've come away numerous times shaking our heads after they've voted against us saying - 'What the hell were they thinking?'"

In July of 1997, the Board finally passed the measure, limiting family scalloping to Wednesday through Sunday. Meanwhile, the heightened interest in the health of the scallop industry prompted a request to Tracy Curley to deliver a brief primer on the scallop, which she did in a letter dated October 16, 1997. She outlined the diet of the scallop, its classification, settlement, growth, life span, and reproductive cycle, noting that environmental conditions affected the animal dramatically. Of key importance is the amount of phytoplankton in the water which, despite being only twenty percent of its food intake (the rest is bacterial detritus), supplies most of its energy. And most of that energy goes to a short but frenzied two month period of gonad development leading up to reproduction, a process which is highly sensitive to water temperature, salinity, day length, and food. In fact, the letter makes clear that the scallop is a gonad dominant animal, the entirety of whose brief existence seems dedicated to the growth and ultimate one-time use (perhaps two, if lucky) of its gonad, achieving in this enviable event a full lifetime of sexual experience, from both the male and female perspective, in one grand climactic act of evolutionary passion.

But at the end of the thesis were two interesting pieces of information. One was Belding's observation that only ten to twenty percent of scallops lived long enough to spawn twice, physiologically degrading after eighteen months. But Tracy also cited a 1963 study linking mortality to environmental stress more than heredity, and a 1978 study linking scallop maturity to water temperature, indicating that slower growing scallops might spat later in the season. This created the possibility of two classes of spat, a June/July spat, and an August/September spat, and the further possibility that the later class of scallops would have a less effective spat the following spring, yet still create a 'nub' growth ring. The term, as far as anyone knows, was coined on Nantucket in response to its seeming exclusive hold on the problem. Tracy thought that the nubs might live to spat more effectively a second season. "The second class spawn may not be productive," she writes, rather innocuously, to close the letter. "Therefore, the majority of scallops for next year's harvest will be a product of the first class spawn. Conversely, if both classes spawn, there will be remarkably more scallops to be caught in the consecutive year."

Nobody, as yet, had any idea where this kernel of an observation would lead. But three weeks later, they would. *The Inquirer and Mirror* splashed two headlines on the front page of the November 6 issue, the opening week. One read: **Scallopers see a slim season ahead**. The other, in italics and accompanied by a graph, said ***Numbers tell the story of a declining scalloping industry***. Inside were three full pages of articles under the larger heading **Scalloping: An Industry in Decline?**, with an even larger subhead saying ***Experts Face a Full Slate of Concerns***. It was comprised of several articles with their own headlines: **Overfishing: Is it a serious problem or a red herring?**; **Development, septic systems, harbor usage have an impact**; **Diehard scallopers unfazed by risk of slim catch**; **A Lifetime spent scalloping**; **Circulation: a pivotal factor**; and **The quality and quantity of eelgrass is a burning question**. They represented a complete and comprehensive summary of the problems facing the Nantucket scallop fishery.

Pandora was fully out of her box now, yet, in reality no one knew precisely what to do. The 1997-98 season was worse than the previous, at 8,000 bushels, and then the 1998-1999 season was the worst of all at 6,800. One could only come to the conclusion that if the trend stayed the same, there might be no fishery left in 1999-2000, or soon thereafter. Worse, no one had ever lost their fishery and then permanently rebuilt it.

saving the scallops

How does Nantucket save its scallop fishery?

The first step is to understand that the challenge is to save the fishery, not just the scallops. Many of the towns that have lost their fisheries have not necessarily lost their scallops. Like the white abalone of California, scallops exist, but not in numbers enough to sustain a commercial or even a recreational fishery. In order to have a commercial fishery, one must maintain a sustainable number of scallops to support the fishing effort, which means understanding that three main factors affect scallops in a fishery: nature, man's effect on their environment, and fishing.

Nature is probably the biggest cause of scallops' cyclicality: Storms affect salinity and turbidity, move scallops out of their beds, blow them ashore, and change the bottom by shoaling over eel grass beds or altering channels. Predators kill their share. Water temperature varies, perhaps affected by global warming, which affects the timing of the spat season and the micro-organisms that baby scallops feed on. Disease to scallops or eel grass can and has caused catastrophic damage.

Man-made environmental changes affect the scallops, in many ways that are only recently being understood or studied. Many of the relevant studies have happened recently, and are in direct response to the decline of the Nantucket Bay Scallop. Some are theories based on observations of other ecosystems. The most devastating problems caused to scallops are loss of eel grass beds associated with algae blooms commonly called brown tides. While not proven to be the cause of the 1931 die-off, empirical evidence from the brown tides of the mid-1980's and mid-1990's in Long Island Sound indicates that they were probably man-generated events from nutrient overloading, much of the blame being placed on agricultural by-products, specifically high concentration of feces from duck-farming. On Nantucket, suspected offenders include point and non-point source pollution from fertilizers, septic systems, and nutrients entering from Nantucket Sound. Warmer water worsens any nutrient problem by encouraging interaction and growth, making the water more turbid, which makes it more difficult for eel grass to grow. It is unclear what effect these changes might have on the microscopic food source of the scallop itself as the ratio of one type of phytoplankton or dinoflagellate to another changes. Global warming, whether natural or man-made, is probably more than just a fisherman's observation: nine of the ten hottest years on record have taken place since 1990. How it affects northern bay scallops is unclear, except that we know their spat season is controlled by water temperature. Petroleum products, grey water, and chemicals which enter the water through boating activity and road runoff , and which may interact with surfactant soaps, probably play a role. While the effect of these altered nutrient and chemical levels may be scientifically unclear or not fully studied, common sense tells one that changes have been made and that the scallops are dying, and therefore inferring some relationship between the two is not unreasonable. And finally, overdredging and dragging of mooring chains, which creates barren circles in eel grass, might be factors.

Fishing practices also affect the health of the scallops. Over the years various things have been discussed or tried regarding fairness to the fishermen and sustainability of the scallop, including daily catch limits, seasonal open limits, limiting areas where fishing is permitted, limiting the number of either family or commercial permits, limiting the hours, limiting the allowable fishing temperature to 28°, studying the the effect of aquaculture and of propagated scallops mingling with the natural set, and regulating the type and size of gear used, including dredge size, weight and number of dredges.

While the nub scallop issue is currently the focus, and is filled with animosity, it is in reality only a small piece of the overall puzzle. In fact, nubs became an issue only after the scallops had already been virtually wiped out. Taking nub scallops probably had little or nothing to do with their initial depletion, but leaving them in the water now is looked upon as one of several significant factors that might aid their recovery. When a definition of a seed that suited all was in force (the state wrestled with a definition from 1907 to 1909, before finally returning to the original 'well-defined, raised growth line' standard), the fishery remained sustainable through many cycles. But since it has never seen a down cycle this long and this severe, traditionally accepted practices may not necessarily still apply.

Fisheries management in general has only recently had to respond to problems related to the commercial extinction of a species, and neither regulators nor fishermen have truly come up to speed in acknowledging this. Most previous issues have dealt with economic or fairness concerns - i.e. how to regulate an open access fishery so that everyone has a fair chance to catch and sell fish. Now, in fishery after fishery, whether scallop or finfish, the problem is no longer about fairness, equity, or market value, but rather about the prospect of simple continued existence of the species.

rules and regulations

For the first hundred years in Massachusetts, few rules existed regarding scalloping, and they changed little. Scallops originally fell under general acts regulating shellfishing, which governed four areas of interest: town regulations; permitting; seizure in vessels; and protection of the fisheries by limiting catch, place, and time of taking. The first mention of scallops was made in 1874, lumping them with other shellfish in regards to Mount Hope Bay. In 1880 they were mentioned again, as the state granted the towns power to regulate their own eel and shellfish industries.

The first regulations specific to scallops came in 1885, when the state set a twenty-five bushel per boat, per day limit, and established a commercial season from September 1 to April 15, shortening it to October 1 through March 31 in 1896. Family season remained year-round. By 1910, seventeen total scallop-related laws had been enacted, mostly regarding permits, season, catch, and town supervision. One law was passed to fund a study of spawning and propagation. In 1910 a limit of ten bushels per person per day was enacted.

1887 saw the first 'seed' scallop legislation, imposing a $20.00 fine per offense, which was upped to $50.00 per offense in 1896. Given the value of $50.00 in 1896, this is an indication of the seriousness with which the state viewed the offense. Curiously, neither act defined a 'seed' scallop, and in 1906 an attempt was made to define seed scallop as being under two inches in diameter. Many mature scallops were smaller than that, so a more detailed and accurate definition was passed in 1907, but was too complicated to be useful, so in 1909 the 'well-defined growth ring' became the standard, and remains so to this day. Scallopers were initially allowed fifteen percent seed, but that was changed the following year to five percent. And in what was clearly an acknowledgement of the significantly lower stakes of the day, Belding concludes the section on seed with the following gentlemanly courtesy:

> "As long as the scallop fishermen refuse to take the immature scallops, there is but little need of the rigid enforcement of the seed scallop law."

Belding concludes his work with a chapter entitled *Methods of Improving the Scallop Industry*. The introductory summary to the chapter reads as follows:

> The scallop fishery presents peculiarities which differentiate it from

other fishing industries, and a knowledge of which is essential in considering its improvement. (1) Protective legislation is principally confined to the seed scallop, or scallop less than one year old. (2) The future welfare lies wholly in the hands of the fishermen and their proper respect for the preservation of the "seed" scallop. (3) There is no great prospect for a wide expansion of the fishery, as there are few ways of artificially increasing the supply; but on the other hand, if the spirit of protective legislation prevails there is but slight danger of serious diminution. (4) The scallop fishery is peculiarly fortunate as, unlike the clamming and quahaug industries, it is unaffected by heavy fishing and needs but minimum care on the part of the fishermen to remain in excellent condition. Thus, while there are few possibilities for increasing the natural supply, there is but slight danger of its permanent extinction.

resistance

A recent incident involving a very minor rule change is both funny and ironic, because it puts into context the opposing lenses through which scallopers and wardens view regulations. Through 1976, scallopers could fish from sunrise to sunset, but in 1977 the rules were changed limiting fishing from 6:30 am to 4:30 pm. Caught and fined on the first day for dropping their dredges early were five fishermen identified in *The Inquirer and Mirror* by last name only - Ames, Harrington, Jelleme, Cabot, and Kotalac. Dave Fronzuto had told me that the law was originally intended as a safety measure and that no one really griped about it. Much later, a scalloper told me that the law was aimed specifically at him, because he used to land after dark with his many pirated scallops. Fronzuto was not yet Harbormaster, nor even an island resident, yet he draws upon the institutional history of his office, and the subsequent value of the rule to his policy, to interpret the intent of the regulation. The scalloper, who is not one of the above-named violators, is convinced that the law takes personal aim at him, a fairly common sentiment. Yet, the five scallopers who were fined clearly must have griped about it, and since none of them were the scalloper who took it so personally, it must have been aimed at curing some larger problem, which might be explained somewhere in the mist of those Selectmen's minutes, if they still exist.

The story, while funny, is telling, because it captures in a nutshell the reality that regulation of any industry, and particularly a fishing industry, is as much about personality and politics as it is about science and knowledge. Fishermen are notoriously independent. They hate to be regulated. Regulators, in this case Tracy and Dave at the local level, backed by the SHAB board and the Division of Marine Fisheries, are convinced that the state of the fishery requires aggressive regulation in order to save it for the commercial fishermen. As the resource has dwindled and the regulations have tightened in response, tempers have flared. Many are growing convinced that the regulators are trying to save it *from* the commercial fishermen. Some believe that the Marine Department is trying to put them out of business, period. Some are offended at the perception that they are being accused of overfishing. Smitty, who a year earlier had been a Fronzuto supporter for the attention that the Marine Department was beginning to focus on the scallop industry, now believes that regulatory practices have gone overboard, and that the commercial fishermen are being made to suffer through Dave Fronzuto's learning curve, without being given any input.

Week after week, as the 2000 season geared up, the nub issue refused to die. There were heated discussions at SHAB meetings, letters to the editor, and countless tense landing inspections. Why? Because most of the old line fishermen believe that the nub scallops will only die anyway, or at best produce a very old scallop the following year which will be less marketable, and because inspecting for them will take more time on the water and potentially result in less money if they cannot reach their limits. On the other side, the Marine Department has not been able to convince many of them of their belief that the only reason that nub scallops are a major issue now is that other stresses on the scallop population, unrelated to fishing practices, have made them necessary to the survival of the harbor. Their view is that, in years past, there had simply been so many more adults and seed remaining at the end of the season that the fishery could sustain a significant overcatch of nubs. The nubs are now an issue because the fishermen, for the first time ever, are relying on them to fill their baskets. The Marine Department's position is that the fishery itself is relying on them also, to maintain a sustainable population, and it intends to rigidly enforce the growth-ring statute. Some of the more ardent scallopers have stated their intentions to form a fishermen's association and seek legal counsel to protect their interests.

acquisition

Can we blame a scalloper for wanting to catch as many scallops as he can? After all, nobody slaps any daily catch limits on lawyers, or businessmen, or athletes, or rock stars. Would one of Nantucket's resident billionaires stand for making only five bushels of money a day? How does one value human endeavor?

The film *Wall Street* coined the phrase 'Greed is Good,' as the motto of unrestrained capitalism. Granted, this somewhat punchy distillation was aimed at theatre audiences, but it still describes the essence of the free enterprise approach to valuing human endeavor: it is worth whatever you can get for it. Few would argue that this beats having some bureaucrat, dictator, or monarch tell you what your labor is worth. And as one famous statesman said 'Capitalism is the worst system ever invented ... except for all the others.' But is 'Better than Communism' really a worthy goal for a nation that considers itself the world's most advanced society?

Film, however, is fiction, and its tenets must be taken in that context. Greed is more easily understood, at least by me, in this real world context: nature always wins. A capitalist world is not the real world - it is an artificial construct that allows people to place varying amounts of insulation between themselves and the laws of nature. The survival quatrain remains unchanged: sustenance, safety, resistance to disease, and procreation.

Man uses his collective resources to outwit nature and stay alive as long as possible. But he knows the eventual outcome is death, and he has learned that acquisition is the best way to postpone it: acquisition of food, clothing, shelter, protection, medicine. Hundreds of millions of years of DNA imprinting created this overwhelming, protozoic impulse to acquire.

To remind myself of the value of mankind's collective conventions, I need only go out my door. It is a typical New England winter night, twenty-five degrees or so. If I were wearing only what nature gave me, I'd be dead by morning. The further one gets away from cities, the more pertinent this knowledge becomes. In cities, car accidents, homicides, and suicides are the main causes of death in healthy people. But in rural areas, the scale is shifted slightly. In farming communities, people get caught in machinery; in fishing communities, people drown. Nature scores an early victory here and there, providing reminders to the DNA to stay sharp.

Greed is not just a desire for wealth, it is deeper. It is part of the genetic need to beat nature; a natural response to the the fact that if you do not beat nature, nature will beat you. It is the striving to understand and control and manipulate and disarm nature in order to co-exist with it. The path is direct. It starts with the most basic of instincts: stay warm, get food, protect yourself, reproduce. With each success, the instinct pushes to make the odds better. Grow food, eliminate predators, cure disease. The next natural step is well underway - understand, control, and manipulate the genetic path.

So far, as the bumper sticker says, nature still bats last. Her finger is still on the switch, but for how long? And as Satchel Paige, that flamethrowing philosopher of the old Negro League once said, 'Be careful what you wish for, you might get it.'

Perhaps, in the last hundred or so generations man has discovered, in his quest for species survival, some benefit to societal sharing, and concepts such as the Comprehensive Plan or fisheries regulations are efforts to reconcile the logic of collective resource management with the emotion of our prehistorically coded greed genes.

Until we resolve this conflict, can we blame a scalloper for wanting to catch as many scallops as he can? Not without being hypocritical. On the other hand, if every scalloper goes out and catches as many scallops as he can today, there will clearly be no scallops left to catch tomorrow. It is entirely possible for no one to be at fault, and everyone to lose anyway. Ask the fishermen in Peconic Bay.

would we miss it?

A WorldWatch 'State of the World' report released on January 16, 2000, estimated that the world has lost 25% of its mammal species, 11% of its birds, and and 34% of its fish species since 1984. Which begs the question - 'Did we need them?'

Would the world go on without the piping plover? Without the spotted owl or the snail darter? The redwood tree? The cod? The buffalo? The whale? The manatee? Lion, tiger, elephant, zebra, cheetah, rhino?

The answer is, sadly, yes. The world would get along just fine without them, just as it is getting along without the many other species we have already eliminated. And truly, the world would also get along just fine without the Nantucket Bay Scallop.

But would we miss it? And how do we explain to our children that we left them a world with half the life on it that our parents left us? And who will we blame for it? We are the richest and most powerful and most influential nation on the planet, and we use the most resources. We have had the biggest impact on its destruction, and we are in a position to have the biggest impact on its salvation. It is hard to imagine twisting logic far enough to find a way to blame it on anyone else. Can we save the whole planet? Not likely. Can we save the Nantucket Bay Scallop fishery? Perhaps.

Afterword

Two years have passed since Rob and I began chronicling what could have been Nantucket's last scallop season. It was not. In fact, that season produced 14,000 bushels of scallops, and another 15,000 in the year 2000-2001, with opening day prices again at $13.00 per pound. In response to the roughly 1,950 recreational permits being issued that year (Dave's cap of 1,800 permits was declared illegal), the family license fee was raised to $25.00, and a non-resident fee of $100.00 was instituted. "Still the best deal on the island," crowed Rob, although his annual quest for 'Number 1' had yet again met with futility.

But circumstances far beyond Nantucket shaped Opening Day, 2001. A beautiful day, calm and nearing seventy degrees, greeted the fleet. Very large scallops were abundant in the harbor, so only a handful of boats fished the smaller Madaket variety - a reversal of our season. But we were stunned by the price being offered at Sayle's Seafood: $8.00 a pound.

Ironically, in light of the tentative success in restoring the fishery, the increased supply may have contributed to the drop in price. There was also a glut of sea scallops available, and a good set in Martha's Vineyard. But the biggest factor was one that no one could have predicted. The attack on the World Trade Center had an enormous psychological and economic impact on Americans, and the wound was still fresh on opening day. Although Eric Ripert, executive chef at New York's Le Bernardin restaurant, told *The Inquirer and Mirror* that the main charm of the Nantucket Bay Scallop is its rarity - "Like Christmas cookies, you only eat them for part of the year - the rest of the time you just look forward to them," - he also noted that demand was drastically down. Tobey Leske offered his insight into the situation: "Osama bin Laden has reached Nantucket. This is a hidden cost of September 11."

As the 2000 census data became available, islanders' everyday observations were statistically borne out. The year-round population had, in fact, increased 58% during the 1990's, from just over 6,000 to nearly 10,000. It is hard to know the population of the island on a typical summer day, but 50,000 would be a good guess. From 1995 to 2000 the number of registered vehicles jumped 46%, to over 14,000. The number of boats in the harbor went from 1,200 to 2,500. The number of houses undoubtedly came close to doubling in ten years.

And the approximate number of scallops landed has dropped from probably 150,000 bushels in 1968, to 117,000 in 1980, to 44,000 in 1990, to 15,000 in 2000. As Tobey Leske had said about his chart: "That is not a cycle, it is a trend." And just as Bobby DeCosta's observation - "I'm not a biologist, just a fisherman, but you take a scallop that's an eighth of an inch big, how much spat can it physically throw out?" - has the unmistakable ring of truth, so has Charlie Sayle's: "Probably got a choice between the tourists and the scallops. There's a lot more money in tourists."

Efforts to restore scallops to the harbor seem to be succeeding, even if there is no consensus on which efforts have been effective and which have not. The health of the harbor has changed, the scallop population has changed, the methods of fishing have changed, and it is even possible that the scallop itself has changed. The belief is virtually unanimous that more than one problem has affected the fishery. It stands to reason, then, that more than one solution will be required to cure it. Nobody yet knows what the magic combination of solutions might be. Not the Division of Marine Fisheries, not Woods Hole, not SHAB, not the Selectmen, not the Marine Department, not a newly-formed fishermen's association, and not even an old-time fisherman. Those solutions will involve some pain and inconvenience to all, from boat owners to home owners to landscapers to taxpayers to regulators to family scallopers to fishermen. But, based on Nantucket's own catch totals and the disastrous experience of almost every other bay scallop fishery in the Northeast, it seems certain that taking no action at all risks the total loss of the fishery. And, like looking for a light switch in a dark room, one has to start somewhere.

One of the values of clinging stubbornly to tradition lies in the strength of heart, and emotion, and will, in keeping something worthy alive. It is a profoundly human characteristic, and one that has prevailed over dispassionate assessment time and time again. It will be necessary to keep the Nantucket scallop fishery alive.

But also, as Bart Giamatti, a beloved professor of Renaissance Studies at Yale, later its president, and finally Commissioner of Baseball until his untimely death, a man who cherished and articulated those institutions' traditions more eloquently than anyone I have ever known, once said: "Great institutions, in order to remain great, must change."

The Nantucket scallop fishery has been a great institution since its humble beginning of a few islanders wading out into the harbor with baskets and pushrakes. It has adapted in the past, as times have changed. Now that the pressures of modern society and greater population have altered the quality of the habitat and decreased the number of the scallops, the island and the fishery must adapt again.

Because yes, we would miss it.